Holt Mathematics

Course 1
Lesson Plans

HOLT, RINEHART AND WINSTON
A Harcourt Education Company
Orlando • Austin • New York • San Diego • Toronto • London

Copyright © by Holt, Rinehart and Winston

All rights reserved. No part of this publication may be reproduced or transmitted in any form or by any means, electronic or mechanical, including photocopy, recording, or any information storage and retrieval system, without permission in writing from the publisher.

Teachers using Holt Mathematics may photocopy complete pages in sufficient quantities for classroom use only and not for resale.

HOLT and the **"Owl Design"** are trademarks licensed to Holt, Rinehart and Winston, registered in the United States of America and/or other jurisdictions.

Printed in the United States of America

If you have received these materials as examination copies free of charge, Holt, Rinehart and Winston retains title to the materials and they may not be resold. Resale of examination copies is strictly prohibited.

Possession of this publication in print format does not entitle users to convert this publication, or any portion of it, into electronic format.

ISBN 0-03-078251-1
6 7 8 9 10   170   10 09 08

Contents

Chapter 1
Lesson 1-1 1
Lesson 1-2 2
Lesson 1-3 3
Lesson 1-4 4
Lesson 1-5 5
Lesson 1-6 6
Lesson 1-7 7

Chapter 2
Lesson 2-1 8
Lesson 2-2 9
Lesson 2-3 10
Lesson 2-4 11
Lesson 2-5 12
Lesson 2-6 13
Lesson 2-7 14
Lesson 2-8 15

Chapter 3
Lesson 3-1 16
Lesson 3-2 17
Lesson 3-3 18
Lesson 3-4 19
Lesson 3-5 20
Lesson 3-6 21
Lesson 3-7 22
Lesson 3-8 23
Lesson 3-9 24

Chapter 4
Lesson 4-1 25
Lesson 4-2 26
Lesson 4-3 27
Lesson 4-4 28
Lesson 4-5 29
Lesson 4-6 30
Lesson 4-7 31
Lesson 4-8 32
Lesson 4-9 33

Chapter 5
Lesson 5-1 34
Lesson 5-2 35
Lesson 5-3 36
Lesson 5-4 37
Lesson 5-5 38
Lesson 5-6 39
Lesson 5-7 40
Lesson 5-8 41
Lesson 5-9 42
Lesson 5-10 43

Chapter 6
Lesson 6-1 44
Lesson 6-2 45
Lesson 6-3 46
Lesson 6-4 47
Lesson 6-5 48
Lesson 6-6 49
Lesson 6-7 50
Lesson 6-8 51
Lesson 6-9 52
Lesson 6-10 53

Chapter 7
Lesson 7-1 54
Lesson 7-2 55
Lesson 7-3 56
Lesson 7-4 57
Lesson 7-5 58
Lesson 7-6 59
Lesson 7-7 60
Lesson 7-8 61
Lesson 7-9 62
Lesson 7-10 63

Chapter 8
Lesson 8-1 64
Lesson 8-2 65
Lesson 8-3 66
Lesson 8-4 67
Lesson 8-5 68
Lesson 8-6 69
Lesson 8-7 70
Lesson 8-8 71
Lesson 8-9 72
Lesson 8-10 73
Lesson 8-11 74

Chapter 9
Lesson 9-1 75
Lesson 9-2 76
Lesson 9-3 77
Lesson 9-4 78
Lesson 9-5 79
Lesson 9-6 80
Lesson 9-7 81
Lesson 9-8 82

Chapter 10
Lesson 10-1 83
Lesson 10-2 84
Lesson 10-3 85
Lesson 10-4 86

Contents

Lesson 10-5 87
Lesson 10-6 88
Lesson 10-7 89
Lesson 10-8 90
Lesson 10-9 91

Chapter 11
Lesson 11-1 92
Lesson 11-2 93
Lesson 11-3 94
Lesson 11-4 95
Lesson 11-5 96
Lesson 11-6 97

Lesson 11-7 98
Lesson 11-8 99
Lesson 11-9 100
Lesson 11-10 101

Chapter 12
Lesson 12-1 102
Lesson 12-2 103
Lesson 12-3 104
Lesson 12-4 105
Lesson 12-5 106
Lesson 12-6 107

Teacher's Name _____ Class _____ Date _____

Lesson Plan 1-1
Comparing and Ordering Whole Numbers pp. 6–9   Day _____

Objective Students compare and order whole numbers using place value or a number line.

> **NCTM Standards:** Understand numbers, ways of representing numbers, relationships among numbers, and number systems; Recognize and apply mathematics in contexts outside of mathematics; Create and use representations to organize, record, and communicate mathematical ideas.

Pacing
☐ 45-minute Classes: 1 day   ☐ 90-minute Classes: 1/2 day   ☐ Other_____

WARM UP
☐ Warm Up TE p. 6 and Daily Transparency 1-1
☐ Problem of the Day TE p. 6 and Daily Transparency 1-1
☐ Countdown to Testing Transparency Week 1

TEACH
☐ Lesson Presentation CD-ROM 1-1
☐ Alternate Opener, Explorations Transparency 1-1, TE p. 6, and Exploration 1-1
☐ Reaching All Learners TE p. 7
☐ Teaching Transparency 1-1
☐ *Hands-On Lab Activities* 1-1
☐ *Know-It Notebook* 1-1

PRACTICE AND APPLY
☐ Example 1: Average: 1–2, 9–11, 21–29, 38, 42–51 Advanced: 1–2, 9–11, 21–29, 40, 42–51
☐ Example 2: Average: 1–39, 42–51 Advanced: 3–51

REACHING ALL LEARNERS – Differentiated Instruction for students with

Developing Knowledge	On-level Knowledge	Advanced Knowledge	English Language Development
☐ Inclusion TE p. 7	☐ Inclusion TE p. 7	☐ Inclusion TE p. 7	☐ Inclusion TE p. 7
☐ Practice A 1-1 CRB	☐ Practice B 1-1 CRB	☐ Practice C 1-1 CRB	☐ Practice A, B, or C 1-1 CRB
☐ Reteach 1-1 CRB	☐ Puzzles, Twisters & Teasers 1-1 CRB	☐ Challenge 1-1 CRB	☐ *Success for ELL* 1-1
☐ Homework Help Online Keyword: MR7 1-1	☐ Homework Help Online Keyword: MR7 1-1	☐ Homework Help Online Keyword: MR7 1-1	☐ Homework Help Online Keyword: MR7 1-1
☐ *Lesson Tutorial Video* 1-1	☐ *Lesson Tutorial Video* 1-1	☐ *Lesson Tutorial Video* 1-1	☐ *Lesson Tutorial Video* 1-1
☐ Reading Strategies 1-1 CRB	☐ Problem Solving 1-1 CRB	☐ Problem Solving 1-1 CRB	☐ Reading Strategies 1-1 CRB
☐ Questioning Strategies p. 1	☐ Visual TE p. 7	☐ Visual TE p. 7	
☐ *IDEA Works!* 1-1			☐ *Multilingual Glossary*

ASSESSMENT
☐ Lesson Quiz, TE p. 9 and DT 1-1   ☐ State-Specific Test Prep Online Keyword: MR7 TestPrep

Copyright © Holt, Rinehart and Winston.
All rights reserved.

Holt Mathematics

Teacher's Name _____ Class _____ Date _____

Lesson Plan 1-2
Estimating with Whole Numbers pp. 10–13 Day _____

Objective Students estimate with whole numbers.

> **NCTM Standards:** Compute fluently and make reasonable estimates.

Pacing
☐ 45-minute Classes: 1 day   ☐ 90-minute Classes: 1/2 day   ☐ Other_____

WARM UP
☐ Warm Up TE p. 10 and Daily Transparency 1-2
☐ Problem of the Day TE p. 10 and Daily Transparency 1-2
☐ Countdown to Testing Transparency Week 1

TEACH
☐ Lesson Presentation CD-ROM 1-2
☐ Alternate Opener, Explorations Transparency 1-2, TE p. 10, and Exploration 1-2
☐ Reaching All Learners TE p. 11
☐ *Know-It Notebook* 1-2

PRACTICE AND APPLY
☐ Example 1: Average: 1–2, 6–9, 25–26, 32–41 Advanced: 1–2, 6–9, 27–28, 32–41
☐ Example 2: Average: 1–3, 6–10, 13–18, 25–26, 32–41 Advanced: 6–10, 13–18, 25–29, 32–41
☐ Example 3: Average: 1–27, 32–41 Advanced: 5–41

REACHING ALL LEARNERS – Differentiated Instruction for students with

Developing Knowledge	On-level Knowledge	Advanced Knowledge	English Language Development
☐ Number Sense TE p. 11	☐ Number Sense TE p. 11	☐ Number Sense TE p. 11	☐ Number Sense TE p. 11
☐ Practice A 1-2 CRB	☐ Practice B 1-2 CRB	☐ Practice C 1-2 CRB	☐ Practice A, B, or C 1-2 CRB
☐ Reteach 1-2 CRB	☐ Puzzles, Twisters & Teasers 1-2 CRB	☐ Challenge 1-2 CRB	☐ *Success for ELL* 1-2
☐ Homework Help Online Keyword: MR7 1-2	☐ Homework Help Online Keyword: MR7 1-2	☐ Homework Help Online Keyword: MR7 1-2	☐ Homework Help Online Keyword: MR7 1-2
☐ *Lesson Tutorial Video* 1-2	☐ *Lesson Tutorial Video* 1-2	☐ *Lesson Tutorial Video* 1-2	☐ *Lesson Tutorial Video* 1-2
☐ Reading Strategies 1-2 CRB	☐ Problem Solving 1-2 CRB	☐ Problem Solving 1-2 CRB	☐ Reading Strategies 1-2 CRB
☐ *Questioning Strategies* pp. 2–3	☐ Communicating Math TE p. 11	☐ Critical Thinking TE p. 11	☐ Lesson Vocabulary SE p. 10
☐ *IDEA Works!* 1-2			☐ *Multilingual Glossary*

ASSESSMENT
☐ Lesson Quiz, TE p. 13 and DT 1-2   ☐ State-Specific Test Prep Online Keyword: MR7 TestPrep

Teacher's Name _____ Class _____ Date _____

Lesson Plan 1-3
Exponents pp. 14–17 Day _____

Objective Students represent numbers by using exponents.

> **NCTM Standards:** Understand numbers, ways of representing numbers, relationships among numbers, and number systems; Compute fluently and make reasonable estimates; Create and use representations to organize, record, and communicate mathematical ideas.

Pacing
☐ 45-minute Classes: 1 day   ☐ 90-minute Classes: 1/2 day   ☐ Other_____

WARM UP
☐ Warm Up TE p. 14 and Daily Transparency 1-3
☐ Problem of the Day TE p. 14 and Daily Transparency 1-3
☐ Countdown to Testing Transparency Week 1

TEACH
☐ Lesson Presentation CD-ROM 1-3
☐ Alternate Opener, Explorations Transparency 1-3, TE p. 14, and Exploration 1-3
☐ Reaching All Learners TE p. 15
☐ Teaching Transparency 1-3
☐ *Technology Lab Activities* 1-3
☐ *Know-It Notebook* 1-3

PRACTICE AND APPLY
☐ Example 1: Average: 1–6, 33–37, 65–71 Advanced: 13–21, 38–42, 65–71
☐ Example 2: Average: 1–12, 33–47, 65–71 Advanced: 13–31, 38–52, 65–71
☐ Example 3: Average: 1–37, 48–71 Advanced: 13–71

REACHING ALL LEARNERS – Differentiated Instruction for students with

Developing Knowledge	On-level Knowledge	Advanced Knowledge	English Language Development
☐ Number Sense TE p. 15	☐ Number Sense TE p. 15	☐ Number Sense TE p. 15	☐ Number Sense TE p. 15
☐ Practice A 1-3 CRB	☐ Practice B 1-3 CRB	☐ Practice C 1-3 CRB	☐ Practice A, B, or C 1-3 CRB
☐ Reteach 1-3 CRB	☐ Puzzles, Twisters & Teasers 1-3 CRB	☐ Challenge 1-3 CRB	☐ *Success for ELL* 1-3
☐ Homework Help Online Keyword: MR7 1-3	☐ Homework Help Online Keyword: MR7 1-3	☐ Homework Help Online Keyword: MR7 1-3	☐ Homework Help Online Keyword: MR7 1-3
☐ *Lesson Tutorial Video* 1-3	☐ *Lesson Tutorial Video* 1-3	☐ *Lesson Tutorial Video* 1-3	☐ *Lesson Tutorial Video* 1-3
☐ Reading Strategies 1-3 CRB	☐ Problem Solving 1-3 CRB	☐ Problem Solving 1-3 CRB	☐ Reading Strategies 1-3 CRB
☐ *Questioning Strategies pp. 4–5*	☐ Auditory TE p. 15	☐ Visual TE p. 15	☐ Lesson Vocabulary SE p. 14
☐ *IDEA Works!* 1-3			☐ *Multilingual Glossary*

ASSESSMENT
☐ Lesson Quiz, TE p. 17 and DT 1-3   ☐ State-Specific Test Prep Online Keyword: MR7 TestPrep

Teacher's Name _____ Class _____ Date _____

Lesson Plan 1-4
Order of Operations pp. 22–25   Day _____

Objective Students use the order of operations.

> **NCTM Standards:** Understand meanings of operations and how they relate to one another; Compute fluently and make reasonable estimates.

Pacing
☐ 45-minute Classes: 1 day   ☐ 90-minute Classes: 1/2 day   ☐ Other _____

WARM UP
☐ Warm Up TE p. 22 and Daily Transparency 1-4
☐ Problem of the Day TE p. 22 and Daily Transparency 1-4
☐ Countdown to Testing Transparency Week 1

TEACH
☐ Lesson Presentation CD-ROM 1-4
☐ Alternate Opener, Explorations Transparency 1-4, TE p. 22, and Exploration 1-4
☐ Reaching All Learners TE p. 23
☐ Teaching Transparency 1-4
☐ *Know-It Notebook* 1-4

PRACTICE AND APPLY
☐ Example 1: Average: 1–3, 8–13, 22–24, 48–58 Advanced: 1–3, 8–13, 22–24, 48–58
☐ Example 2: Average: 1–6, 8–19, 22–37, 43–44, 48–58 Advanced: 8–19, 22–41, 43–46, 48–58
☐ Example 3: Average: 1–37, 42–45, 48–58 Advanced: 7–58

REACHING ALL LEARNERS – Differentiated Instruction for students with

Developing Knowledge	On-level Knowledge	Advanced Knowledge	English Language Development
☐ Cooperative Learning TE p. 23	☐ Cooperative Learning TE p. 23	☐ Cooperative Learning TE p. 23	☐ Cooperative Learning TE p. 23
☐ Practice A 1-4 CRB	☐ Practice B 1-4 CRB	☐ Practice C 1-4 CRB	☐ Practice A, B, or C 1-4 CRB
☐ Reteach 1-4 CRB	☐ Puzzles, Twisters & Teasers 1-4 CRB	☐ Challenge 1-4 CRB	☐ *Success for ELL* 1-4
☐ Homework Help Online Keyword: MR7 1-4	☐ Homework Help Online Keyword: MR7 1-4	☐ Homework Help Online Keyword: MR7 1-4	☐ Homework Help Online Keyword: MR7 1-4
☐ *Lesson Tutorial Video* 1-4	☐ *Lesson Tutorial Video* 1-4	☐ *Lesson Tutorial Video* 1-4	☐ *Lesson Tutorial Video* 1-4
☐ Reading Strategies 1-4 CRB	☐ Problem Solving 1-4 CRB	☐ Problem Solving 1-4 CRB	☐ Reading Strategies 1-4 CRB
☐ *Questioning Strategies* pp. 6–7			☐ Lesson Vocabulary SE p. 22
☐ *IDEA Works!* 1-4			☐ *Multilingual Glossary*

ASSESSMENT
☐ Lesson Quiz, TE p. 25 and DT 1-4   ☐ State-Specific Test Prep Online Keyword: MR7 TestPrep

Teacher's Name _____ Class _____ Date _____

Lesson Plan 1-5
Mental Math pp. 26–29   Day _____

Objective Students use number properties to compute mentally.

> **NCTM Standards:** Understand meanings of operations and how they relate to one another.

Pacing
☐ 45-minute Classes: 1 day   ☐ 90-minute Classes: 1/2 day   ☐ Other _____

WARM UP
☐ Warm Up TE p. 26 and Daily Transparency 1-5
☐ Problem of the Day TE p. 26 and Daily Transparency 1-5
☐ Countdown to Testing Transparency Week 2

TEACH
☐ Lesson Presentation CD-ROM 1-5
☐ Alternate Opener, Explorations Transparency 1-5, TE p. 26, and Exploration 1-5
☐ Reaching All Learners TE p. 27
☐ Teaching Transparency 1-5
☐ *Know-It Notebook* 1-5

PRACTICE AND APPLY
☐ Example 1: Average: 1–6, 15–20, 56, 62–69 Advanced: 1–6, 15–20, 58, 62–69
☐ Example 2: Average: 1–42, 62–69 Advanced: 1–39, 43–50, 55–69

REACHING ALL LEARNERS – Differentiated Instruction for students with

Developing Knowledge	On-level Knowledge	Advanced Knowledge	English Language Development
☐ Kinesthetic Experience TE p. 27	☐ Kinesthetic Experience TE p. 27	☐ Kinesthetic Experience TE p. 27	☐ Kinesthetic Experience TE p. 27
☐ Practice A 1-5 CRB	☐ Practice B 1-5 CRB	☐ Practice C 1-5 CRB	☐ Practice A, B, or C 1-5 CRB
☐ Reteach 1-5 CRB	☐ Puzzles, Twisters & Teasers 1-5 CRB	☐ Challenge 1-5 CRB	☐ *Success for ELL* 1-5
☐ Homework Help Online Keyword: MR7 1-5	☐ Homework Help Online Keyword: MR7 1-5	☐ Homework Help Online Keyword: MR7 1-5	☐ Homework Help Online Keyword: MR7 1-5
☐ *Lesson Tutorial Video* 1-5	☐ *Lesson Tutorial Video* 1-5	☐ *Lesson Tutorial Video* 1-5	☐ *Lesson Tutorial Video* 1-5
☐ Reading Strategies 1-5 CRB	☐ Problem Solving 1-5 CRB	☐ Problem Solving 1-5 CRB	☐ Reading Strategies 1-5 CRB
☐ *Questioning Strategies* pp. 8–9			☐ Lesson Vocabulary SE p. 26
☐ *IDEA Works!* 1-5			☐ *Multilingual Glossary*

ASSESSMENT
☐ Lesson Quiz, TE p. 29 and DT 1-5   ☐ State-Specific Test Prep Online Keyword: MR7 TestPrep

Teacher's Name _____ Class _____ Date _____

Lesson Plan 1-6
Choose the Method of Computation pp. 30–32 Day ____

Objective Students choose an appropriate method of computation and justify their choice.

> **NCTM Standards:** Compute fluently and make reasonable estimates; Recognize reasoning and proof as fundamental aspects of mathematics.

Pacing
☐ 45-minute Classes: 1 day ☐ 90-minute Classes: 1/2 day ☐ Other_____

WARM UP
☐ Warm Up TE p. 30 and Daily Transparency 1-6
☐ Problem of the Day TE p. 30 and Daily Transparency 1-6
☐ Countdown to Testing Transparency Week 2

TEACH
☐ Lesson Presentation CD-ROM 1-6
☐ Alternate Opener, Explorations Transparency 1-6, TE p. 30, and Exploration 1-6
☐ Reaching All Learners TE p. 31
☐ *Hands-On Lab Activities* 1-6
☐ *Know-It Notebook* 1-6

PRACTICE AND APPLY
☐ Example 1: Average: 1–15, 18–26 Advanced: 1–6, 9–26

REACHING ALL LEARNERS – Differentiated Instruction for students with

Developing Knowledge	On-level Knowledge	Advanced Knowledge	English Language Development
☐ Cooperative Learning TE p. 31	☐ Cooperative Learning TE p. 31	☐ Cooperative Learning TE p. 31	☐ Cooperative Learning TE p. 31
☐ Practice A 1-6 CRB	☐ Practice B 1-6 CRB	☐ Practice C 1-6 CRB	☐ Practice A, B, or C 1-6 CRB
☐ Reteach 1-6 CRB	☐ Puzzles, Twisters & Teasers 1-6 CRB	☐ Challenge 1-6 CRB	☐ *Success for ELL* 1-6
☐ Homework Help Online Keyword: MR7 1-6	☐ Homework Help Online Keyword: MR7 1-6	☐ Homework Help Online Keyword: MR7 1-6	☐ Homework Help Online Keyword: MR7 1-6
☐ *Lesson Tutorial Video* 1-6	☐ *Lesson Tutorial Video* 1-6	☐ *Lesson Tutorial Video* 1-6	☐ *Lesson Tutorial Video* 1-6
☐ Reading Strategies 1-6 CRB	☐ Problem Solving 1-6 CRB	☐ Problem Solving 1-6 CRB	☐ Reading Strategies 1-6 CRB
☐ *Questioning Strategies* pp. 10–11			
☐ *IDEA Works!* 1-6			☐ *Multilingual Glossary*

ASSESSMENT
☐ Lesson Quiz, TE p. 32 and DT 1-6   ☐ State-Specific Test Prep Online Keyword: MR7 TestPrep

Teacher's Name _____ Class _____ Date _____

Lesson Plan 1-7
Patterns and Sequences pp. 33–36   Day _____

Objective Students find, recognize, describe, and extend patterns in sequences.

> **NCTM Standards:** Understand patterns, relations, and functions; Recognize reasoning and proof as fundamental aspects of mathematics; Make and investigate mathematical conjectures; Select and use various types of reasoning and methods of proof.

Pacing
☐ 45-minute Classes: 1 day   ☐ 90-minute Classes: 1/2 day   ☐ Other _____

WARM UP
☐ Warm Up TE p. 33 and Daily Transparency 1-7
☐ Problem of the Day TE p. 33 and Daily Transparency 1-7
☐ Countdown to Testing Transparency Week 2

TEACH
☐ Lesson Presentation CD-ROM 1-7
☐ Alternate Opener, Explorations Transparency 1-7, TE p. 33, and Exploration 1-7
☐ Reaching All Learners TE p. 34
☐ *Hands-On Lab Activities* 1-7
☐ *Know-It Notebook* 1-7

PRACTICE AND APPLY
☐ Example 1: Average: 1–4, 8–11, 16–18, 27–32 Advanced: 8–11, 16–22, 27–32
☐ Example 2: Average: 1–23, 27–32 Advanced: 5–32

REACHING ALL LEARNERS – Differentiated Instruction for students with

Developing Knowledge	On-level Knowledge	Advanced Knowledge	English Language Development
☐ Multiple Representations TE p. 34	☐ Multiple Representations TE p. 34	☐ Multiple Representations TE p. 34	☐ Multiple Representations TE p. 34
☐ Practice A 1-7 CRB	☐ Practice B 1-7 CRB	☐ Practice C 1-7 CRB	☐ Practice A, B, or C 1-7 CRB
☐ Reteach 1-7 CRB	☐ Puzzles, Twisters & Teasers 1-7 CRB	☐ Challenge 1-7 CRB	☐ *Success for ELL* 1-7
☐ Homework Help Online Keyword: MR7 1-7	☐ Homework Help Online Keyword: MR7 1-7	☐ Homework Help Online Keyword: MR7 1-7	☐ Homework Help Online Keyword: MR7 1-7
☐ *Lesson Tutorial Video* 1-7	☐ *Lesson Tutorial Video* 1-7	☐ *Lesson Tutorial Video* 1-7	☐ *Lesson Tutorial Video* 1-7
☐ Reading Strategies 1-7 CRB	☐ Problem Solving 1-7 CRB	☐ Problem Solving 1-7 CRB	☐ Reading Strategies 1-7 CRB
☐ *Questioning Strategies* pp. 12–13			☐ Lesson Vocabulary SE p. 33
☐ *IDEA Works!* 1-7			☐ *Multilingual Glossary*

ASSESSMENT
☐ Lesson Quiz, TE p. 36, and DT 1-7   ☐ State-Specific Test Prep Online Keyword: MR7 TestPrep

Teacher's Name _____ Class _____ Date _____

Lesson Plan 2-1
Variables and Expressions pp. 54–57   Day _____

Objective Students identify and evaluate expressions.

> **NCTM Standards:** Represent and analyze mathematical situations and structures using algebraic symbols.

Pacing
☐ 45-minute Classes: 1 day   ☐ 90-minute Classes: 1/2 day   ☐ Other_____

WARM UP
☐ Warm Up TE p. 54 and Daily Transparency 2-1
☐ Problem of the Day TE p. 54 and Daily Transparency 2-1
☐ Countdown to Testing Transparency Week 3

TEACH
☐ Lesson Presentation CD-ROM 2-1
☐ Alternate Opener, Explorations Transparency 2-1, TE p. 54, and Exploration 2-1
☐ Reaching All Learners TE p. 55
☐ *Technology Lab Activities* 2-1
☐ *Know-It Notebook* 2-1

PRACTICE AND APPLY
☐ Example 1: Average: 1–2, 4–5, 11–22, 24, 28–33 Advanced: 1–2, 4–5, 11–22, 27–33
☐ Example 2: Average: 1–22, 28–33 Advanced: 1–6, 11–33

REACHING ALL LEARNERS – Differentiated Instruction for students with

Developing Knowledge	On-level Knowledge	Advanced Knowledge	English Language Development
☐ Cooperative Learning TE p. 55	☐ Cooperative Learning TE p. 55	☐ Cooperative Learning TE p. 55	☐ Cooperative Learning TE p. 55
☐ Practice A 2-1 CRB	☐ Practice B 2-1 CRB	☐ Practice C 2-1 CRB	☐ Practice A, B, or C 2-1 CRB
☐ Reteach 2-1 CRB	☐ Puzzles, Twisters & Teasers 2-1 CRB	☐ Challenge 2-1 CRB	☐ *Success for ELL* 2-1
☐ Homework Help Online Keyword: MR7 2-1	☐ Homework Help Online Keyword: MR7 2-1	☐ Homework Help Online Keyword: MR7 2-1	☐ Homework Help Online Keyword: MR7 2-1
☐ *Lesson Tutorial Video* 2-1	☐ *Lesson Tutorial Video* 2-1	☐ *Lesson Tutorial Video* 2-1	☐ *Lesson Tutorial Video* 2-1
☐ Reading Strategies 2-1 CRB	☐ Problem Solving 2-1 CRB	☐ Problem Solving 2-1 CRB	☐ Reading Strategies 2-1 CRB
☐ *Questioning Strategies* pp. 14–15	☐ Multiple Representations TE p. 55	☐ Multiple Representations TE p. 55	☐ Lesson Vocabulary SE p. 54
☐ *IDEA Works!* 2-1			☐ *Multilingual Glossary*

ASSESSMENT
☐ Lesson Quiz, TE p. 57 and DT 2-1   ☐ State-Specific Test Prep Online Keyword: MR7 TestPrep

Teacher's Name _____ Class _____ Date _____

Lesson Plan 2-2
Translate Between Words and Math pp. 58–61 Day ____

Objective Students translate between words and math.

> **NCTM Standards:** Represent and analyze mathematical situations and structures using algebraic symbols; Recognize reasoning and proof as fundamental aspects of mathematics; Create and use representations to organize, record, and communicate mathematical ideas.

Pacing
☐ 45-minute Classes: 1 day   ☐ 90-minute Classes: 1/2 day   ☐ Other_____

WARM UP
☐ Warm Up TE p. 58 and Daily Transparency 2-2
☐ Problem of the Day TE p. 58 and Daily Transparency 2-2
☐ Countdown to Testing Transparency Week 3

TEACH
☐ Lesson Presentation CD-ROM 2-2
☐ Alternate Opener, Explorations Transparency 2-2, TE p. 58, and Exploration 2-2
☐ Reaching All Learners TE p. 59
☐ Teaching Transparency 2-2
☐ *Know-It Notebook* 2-2

PRACTICE AND APPLY
☐ Example 1: Average: 1, 12–13, 34, 36–37, 41–49 Advanced: 1, 12–13, 35, 38–39, 41–49
☐ Example 2: Average: 1–7, 12–19, 28–37, 41–49 Advanced: 1–7, 12–19, 28–34, 37–39, 41–49
☐ Example 3: Average: 1–37, 41–49 Advanced: 1–34, 38–49

REACHING ALL LEARNERS – Differentiated Instruction for students with

Developing Knowledge	On-level Knowledge	Advanced Knowledge	English Language Development
☐ Home Connection TE p. 59	☐ Home Connection TE p. 59	☐ Home Connection TE p. 59	☐ Home Connection TE p. 59
☐ Practice A 2-2 CRB	☐ Practice B 2-2 CRB	☐ Practice C 2-2 CRB	☐ Practice A, B, or C 2-2 CRB
☐ Reteach 2-2 CRB	☐ Puzzles, Twisters & Teasers 2-2 CRB	☐ Challenge 2-2 CRB	☐ *Success for ELL* 2-2
☐ Homework Help Online Keyword: MR7 2-2	☐ Homework Help Online Keyword: MR7 2-2	☐ Homework Help Online Keyword: MR7 2-2	☐ Homework Help Online Keyword: MR7 2-2
☐ *Lesson Tutorial Video* 2-2	☐ *Lesson Tutorial Video* 2-2	☐ *Lesson Tutorial Video* 2-2	☐ *Lesson Tutorial Video* 2-2
☐ Reading Strategies 2-2 CRB	☐ Problem Solving 2-2 CRB	☐ Problem Solving 2-2 CRB	☐ Reading Strategies 2-2 CRB
☐ *Questioning Strategies* pp. 16–17	☐ Communicating Math TE p. 59	☐ Communicating Math TE p. 59	
☐ *IDEA Works!* 2-2			☐ *Multilingual Glossary*

ASSESSMENT
☐ Lesson Quiz, TE p. 61 and DT 2-2   ☐ State-Specific Test Prep Online Keyword: MR7 TestPrep

Teacher's Name _____ Class _____ Date _____

Lesson Plan 2-3
Translating Between Tables and Expressions pp. 62–65   Day _____

Objective Students write expressions for tables and sequences.

> **NCTM Standards:** Represent and analyze mathematical situations and structures using algebraic symbols.

Pacing
☐ 45-minute Classes: 1 day   ☐ 90-minute Classes: 1/2 day   ☐ Other _____

WARM UP
☐ Warm Up TE p. 62 and Daily Transparency 2-3
☐ Problem of the Day TE p. 62 and Daily Transparency 2-3
☐ Countdown to Testing Transparency Week 3

TEACH
☐ Lesson Presentation CD-ROM 2-3
☐ Alternate Opener, Explorations Transparency 2-3, TE p. 62, and Exploration 2-3
☐ Reaching All Learners TE p. 63
☐ *Know-It Notebook* 2-3

PRACTICE AND APPLY
☐ Example 1: Average: 1, 4–5, 9–14, 17–26 Advanced: 1, 4–5, 9–11, 13–15, 17–26
☐ Example 2: Average: 1–2, 4–7, 9–14, 17–26 Advanced: 1–2, 4–7, 10–15, 17–26
☐ Example 3: Average: 1–14, 17–26 Advanced: 1–8, 12–26

REACHING ALL LEARNERS – Differentiated Instruction for students with

Developing Knowledge	On-level Knowledge	Advanced Knowledge	English Language Development
☐ Cognitive Strategies TE p. 63	☐ Cognitive Strategies TE p. 63	☐ Cognitive Strategies TE p. 63	☐ Cognitive Strategies TE p. 63
☐ Practice A 2-3 CRB	☐ Practice B 2-3 CRB	☐ Practice C 2-3 CRB	☐ Practice A, B, or C 2-3 CRB
☐ Reteach 2-3 CRB	☐ Puzzles, Twisters & Teasers 2-3 CRB	☐ Challenge 2-3 CRB	☐ *Success for ELL* 2-3
☐ Homework Help Online Keyword: MR7 2-3	☐ Homework Help Online Keyword: MR7 2-3	☐ Homework Help Online Keyword: MR7 2-3	☐ Homework Help Online Keyword: MR7 2-3
☐ *Lesson Tutorial Video* 2-3	☐ *Lesson Tutorial Video* 2-3	☐ *Lesson Tutorial Video* 2-3	☐ *Lesson Tutorial Video* 2-3
☐ Reading Strategies 2-3 CRB	☐ Problem Solving 2-3 CRB	☐ Problem Solving 2-3 CRB	☐ Reading Strategies 2-3 CRB
☐ *Questioning Strategies* pp. 18–19			
☐ *IDEA Works!* 2-3			☐ *Multilingual Glossary*

ASSESSMENT
☐ Lesson Quiz, TE p. 65 and DT 2-3   ☐ State-Specific Test Prep Online Keyword: MR7 TestPrep

Teacher's Name _____ Class _____ Date _____

Lesson Plan 2-4
Equations and Their Solutions pp. 70–73   Day _____

Objective Students determine whether a number is a solution of an equation.

> **NCTM Standards:** Represent and analyze mathematical situations and structures using algebraic symbols; Use representations to model and interpret physical, social, and mathematical phenomena.

Pacing
- ☐ 45-minute Classes: 1 day
- ☐ 90-minute Classes: 1/2 day
- ☐ Other_____

WARM UP
- ☐ Warm Up TE p. 70 and Daily Transparency 2-4
- ☐ Problem of the Day TE p. 70 and Daily Transparency 2-4
- ☐ Countdown to Testing Transparency Week 3

TEACH
- ☐ Lesson Presentation CD-ROM 2-4
- ☐ Alternate Opener, Explorations Transparency 2-4, TE p. 70, and Exploration 2-4
- ☐ Reaching All Learners TE p. 71
- ☐ Teaching Transparency 2-4
- ☐ *Know-It Notebook* 2-4

PRACTICE AND APPLY
- ☐ Example 1: Average: 1–6, 8–19, 22–41, 46–52 Advanced: 1–6, 8–19, 26–44, 46–52
- ☐ Example 2: Average: 1–34, 36–41, 46–52 Advanced: 1–35, 42–52

REACHING ALL LEARNERS – Differentiated Instruction for students with

Developing Knowledge	On-level Knowledge	Advanced Knowledge	English Language Development
☐ Inclusion TE p. 71	☐ Critical Thinking TE p. 71	☐ Critical Thinking TE p. 71	☐ Critical Thinking TE p. 71
☐ Practice A 2-4 CRB	☐ Practice B 2-4 CRB	☐ Practice C 2-4 CRB	☐ Practice A, B, or C 2-4 CRB
☐ Reteach 2-4 CRB	☐ Puzzles, Twisters & Teasers 2-4 CRB	☐ Challenge 2-4 CRB	☐ *Success for ELL* 2-4
☐ Homework Help Online Keyword: MR7 2-4	☐ Homework Help Online Keyword: MR7 2-4	☐ Homework Help Online Keyword: MR7 2-4	☐ Homework Help Online Keyword: MR7 2-4
☐ *Lesson Tutorial Video* 2-4	☐ *Lesson Tutorial Video* 2-4	☐ *Lesson Tutorial Video* 2-4	☐ *Lesson Tutorial Video* 2-4
☐ Reading Strategies 2-4 CRB	☐ Problem Solving 2-4 CRB	☐ Problem Solving 2-4 CRB	☐ Reading Strategies 2-4 CRB
☐ *Questioning Strategies* pp. 20–21			☐ *Lesson Vocabulary* SE p. 70
☐ *IDEA Works!* 2-4			☐ *Multilingual Glossary*

ASSESSMENT
- ☐ Lesson Quiz, TE p. 73 and DT 2-4
- ☐ State-Specific Test Prep Online Keyword: MR7 TestPrep

Teacher's Name _____ Class _____ Date _____

Lesson Plan 2-5
Addition Equations pp. 74–77   Day _____

Objective Students solve whole number addition equations.

> **NCTM Standards:** Understand meanings of operations and how they relate to one another; Represent and analyze mathematical situations and structures using algebraic symbols; Use mathematical models to represent and understand quantitative relationships.

Pacing
☐ 45-minute Classes: 1 day   ☐ 90-minute Classes: 1/2 day   ☐ Other _____

WARM UP
☐ Warm Up TE p. 74 and Daily Transparency 2-5
☐ Problem of the Day TE p. 74 and Daily Transparency 2-5
☐ Countdown to Testing Transparency Week 4

TEACH
☐ Lesson Presentation CD-ROM 2-5
☐ Alternate Opener, Explorations Transparency 2-5, TE p. 74, and Exploration 2-5
☐ Reaching All Learners TE p. 75
☐ Teaching Transparency 2-5
☐ *Hands-On Lab Activities* 2-5
☐ *Know-It Notebook* 2-5

PRACTICE AND APPLY
☐ Example 1: Average: 1–6, 8–16, 18–29, 37–43 Advanced: 1–6, 8–16, 21–29, 30–32, 37–43
☐ Example 2: Average: 1–31, 37–43 Advanced: 1–26, 32–43

REACHING ALL LEARNERS – Differentiated Instruction for students with

Developing Knowledge	On-level Knowledge	Advanced Knowledge	English Language Development
☐ Concrete Manipulatives TE p. 75	☐ Concrete Manipulatives TE p. 75	☐ Concrete Manipulatives TE p. 75	☐ Concrete Manipulatives TE p. 75
☐ Practice A 2-5 CRB	☐ Practice B 2-5 CRB	☐ Practice C 2-5 CRB	☐ Practice A, B, or C 2-5 CRB
☐ Reteach 2-5 CRB	☐ Puzzles, Twisters & Teasers 2-5 CRB	☐ Challenge 2-5 CRB	☐ *Success for ELL* 2-5
☐ Homework Help Online Keyword: MR7 2-5	☐ Homework Help Online Keyword: MR7 2-5	☐ Homework Help Online Keyword: MR7 2-5	☐ Homework Help Online Keyword: MR7 2-5
☐ *Lesson Tutorial Video* 2-5	☐ *Lesson Tutorial Video* 2-5	☐ *Lesson Tutorial Video* 2-5	☐ *Lesson Tutorial Video* 2-5
☐ Reading Strategies 2-5 CRB	☐ Problem Solving 2-5 CRB	☐ Problem Solving 2-5 CRB	☐ Reading Strategies 2-5 CRB
☐ *Questioning Strategies* pp. 22–23	☐ Modeling TE p. 75	☐ Modeling TE p. 75	
☐ *IDEA Works!* 2-5			☐ *Multilingual Glossary*

ASSESSMENT
☐ Lesson Quiz, TE p. 77 and DT 2-5   ☐ State-Specific Test Prep Online Keyword: MR7 TestPrep

Teacher's Name _____ Class _____ Date _____

Lesson Plan 2-6
Subtraction Equations pp. 78–80   Day _____

Objective Students solve whole-number subtraction equations.

> **NCTM Standards:** Understand meanings of operations and how they relate to one another; Represent and analyze mathematical situations and structures using algebraic symbols; Use mathematical models to represent and understand quantitative relationships.

Pacing
☐ 45-minute Classes: 1 day   ☐ 90-minute Classes: 1/2 day   ☐ Other _____

WARM UP
☐ Warm Up TE p. 78 and Daily Transparency 2-6
☐ Problem of the Day TE p. 78 and Daily Transparency 2-6
☐ Countdown to Testing Transparency Week 4

TEACH
☐ Lesson Presentation CD-ROM 2-6
☐ Alternate Opener, Explorations Transparency 2-6, TE p. 78, and Exploration 2-6
☐ Reaching All Learners TE p. 79
☐ *Know-It Notebook* 2-6

PRACTICE AND APPLY
☐ Example 1: Average: 1–29, 33–41 Advanced: 7–41

REACHING ALL LEARNERS – Differentiated Instruction for students with

Developing Knowledge	On-level Knowledge	Advanced Knowledge	English Language Development
☐ Inclusion TE p. 79	☐ Inclusion TE p. 79	☐ Inclusion TE p. 79	☐ Inclusion TE p. 79
☐ Practice A 2-6 CRB	☐ Practice B 2-6 CRB	☐ Practice C 2-6 CRB	☐ Practice A, B, or C 2-6 CRB
☐ Reteach 2-6 CRB	☐ Puzzles, Twisters & Teasers 2-6 CRB	☐ Challenge 2-6 CRB	☐ *Success for ELL* 2-6
☐ Homework Help Online Keyword: MR7 2-6	☐ Homework Help Online Keyword: MR7 2-6	☐ Homework Help Online Keyword: MR7 2-6	☐ Homework Help Online Keyword: MR7 2-6
☐ *Lesson Tutorial Video* 2-6	☐ *Lesson Tutorial Video* 2-6	☐ *Lesson Tutorial Video* 2-6	☐ *Lesson Tutorial Video* 2-6
☐ Reading Strategies 2-6 CRB	☐ Problem Solving 2-6 CRB	☐ Problem Solving 2-6 CRB	☐ Reading Strategies 2-6 CRB
☐ Questioning Strategies pp. 24–25			
☐ *IDEA Works!* 2-6			☐ *Multilingual Glossary*

ASSESSMENT
☐ Lesson Quiz, TE p. 80 and DT 2-6   ☐ State-Specific Test Prep Online Keyword: MR7 TestPrep

Teacher's Name _____ Class _____ Date _____

Lesson Plan 2-7
Multiplication Equations pp. 81–84   Day _____

Objective Students solve whole-number multiplication equations.

> **NCTM Standards:** Understand meanings of operations and how they relate to one another; Represent and analyze mathematical situations and structures using algebraic symbols; Use mathematical models to represent and understand quantitative relationships.

Pacing
☐ 45-minute Classes: 1 day   ☐ 90-minute Classes: 1/2 day   ☐ Other_____

WARM UP
☐ Warm Up TE p. 81 and Daily Transparency 2-7
☐ Problem of the Day TE p. 81 and Daily Transparency 2-7
☐ Countdown to Testing Transparency Week 4

TEACH
☐ Lesson Presentation CD-ROM 2-7
☐ Alternate Opener, Explorations Transparency 2-7, TE p. 81, and Exploration 2-7
☐ Reaching All Learners TE p. 82
☐ *Hands-On Lab Activities* 2-7
☐ *Know-It Notebook* 2-7

PRACTICE AND APPLY
☐ Example 1: Average: 1–6, 8–16, 37, 43–50 Advanced: 1–6, 8–16, 41, 43–50
☐ Example 2: Average: 1–40, 43–50 Advanced: 3–50

REACHING ALL LEARNERS – Differentiated Instruction for students with

Developing Knowledge	On-level Knowledge	Advanced Knowledge	English Language Development
☐ Number Sense TE p. 82	☐ Number Sense TE p. 82	☐ Number Sense TE p. 82	☐ Number Sense TE p. 82
☐ Practice A 2-7 CRB	☐ Practice B 2-7 CRB	☐ Practice C 2-7 CRB	☐ Practice A, B, or C 2-7 CRB
☐ Reteach 2-7 CRB	☐ Puzzles, Twisters & Teasers 2-7 CRB	☐ Challenge 2-7 CRB	☐ *Success for ELL* 2-7
☐ Homework Help Online Keyword: MR7 2-7	☐ Homework Help Online Keyword: MR7 2-7	☐ Homework Help Online Keyword: MR7 2-7	☐ Homework Help Online Keyword: MR7 2-7
☐ *Lesson Tutorial Video* 2-7	☐ *Lesson Tutorial Video* 2-7	☐ *Lesson Tutorial Video* 2-7	☐ *Lesson Tutorial Video* 2-7
☐ Reading Strategies 2-7 CRB	☐ Problem Solving 2-7 CRB	☐ Problem Solving 2-7 CRB	☐ Reading Strategies 2-7 CRB
☐ *Questioning Strategies* pp. 26–27	☐ Cognitive Strategies TE p. 82	☐ Cognitive Strategies TE p. 82	
☐ *IDEA Works!* 2-7			☐ *Multilingual Glossary*

ASSESSMENT
☐ Lesson Quiz, TE p. 84 and DT 2-7   ☐ State-Specific Test Prep Online Keyword: MR7 TestPrep

Teacher's Name _____ Class _____ Date _____

Lesson Plan 2-8
Division Equations pp. 85–87   Day _____

Objective Students solve whole-number division equations.

> **NCTM Standards:** Understand meanings of operations and how they relate to one another; Represent and analyze mathematical situations and structures using algebraic symbols; Use mathematical models to represent and understand quantitative relationships.

Pacing
- [ ] 45-minute Classes: 1 day
- [ ] 90-minute Classes: 1/2 day
- [ ] Other _____

WARM UP
- [ ] Warm Up TE p. 85 and Daily Transparency 2-8
- [ ] Problem of the Day TE p. 85 and Daily Transparency 2-8
- [ ] Countdown to Testing Transparency Week 4

TEACH
- [ ] Lesson Presentation CD-ROM 2-8
- [ ] Alternate Opener, Explorations Transparency 2-8, TE p. 85, and Exploration 2-8
- [ ] Reaching All Learners TE p. 86
- [ ] *Know-It Notebook* 2-8

PRACTICE AND APPLY
- [ ] Example 1: Average: 1–8, 10–17, 19–26, 32–40 Advanced: 1–8, 10–17, 19–25, 28, 32–40
- [ ] Example 2: Average: 1–28, 32–40 Advanced: 5–25, 27–40

REACHING ALL LEARNERS – Differentiated Instruction for students with

Developing Knowledge	On-level Knowledge	Advanced Knowledge	English Language Development
☐ Kinesthetic Experience TE p. 86	☐ Kinesthetic Experience TE p. 86	☐ Kinesthetic Experience TE p. 86	☐ Kinesthetic Experience TE p. 86
☐ Practice A 2-8 CRB	☐ Practice B 2-8 CRB	☐ Practice C 2-8 CRB	☐ Practice A, B, or C 2-8 CRB
☐ Reteach 2-8 CRB	☐ Puzzles, Twisters & Teasers 2-8 CRB	☐ Challenge 2-8 CRB	☐ *Success for ELL* 2-8
☐ Homework Help Online Keyword: MR7 2-8	☐ Homework Help Online Keyword: MR7 2-8	☐ Homework Help Online Keyword: MR7 2-8	☐ Homework Help Online Keyword: MR7 2-8
☐ *Lesson Tutorial Video* 2-8	☐ *Lesson Tutorial Video* 2-8	☐ *Lesson Tutorial Video* 2-8	☐ *Lesson Tutorial Video* 2-8
☐ Reading Strategies 2-8 CRB	☐ Problem Solving 2-8 CRB	☐ Problem Solving 2-8 CRB	☐ Reading Strategies 2-8 CRB
☐ *Questioning Strategies* pp. 28–29			
☐ *IDEA Works!* 2-8			☐ *Multilingual Glossary*

ASSESSMENT
- [ ] Lesson Quiz, TE p. 87 and DT 2-8   ☐ State-Specific Test Prep Online Keyword: MR7 TestPrep

Teacher's Name _____ Class _____ Date _____

Lesson Plan 3-1
Representing, Comparing, and Ordering Decimals pp. 108–111   Day _____

Objective Students write, compare, and order decimals using place values and number lines.

> **NCTM Standards:** Understand numbers, ways of representing numbers, relationships among numbers, and number systems; Recognize and apply mathematics in contexts outside of mathematics; Create and use representations to organize, record, and communicate mathematical ideas.

Pacing
☐ 45-minute Classes: 1 day   ☐ 90-minute Classes: 1/2 day   ☐ Other_____

WARM UP
☐ Warm Up TE p. 108 and Daily Transparency 3-1
☐ Problem of the Day TE p. 108 and Daily Transparency 3-1
☐ Countdown to Testing Transparency Week 5

TEACH
☐ Lesson Presentation CD-ROM 3-1
☐ Alternate Opener, Explorations Transparency 3-1, TE p. 108, and Exploration 3-1
☐ Reaching All Learners TE p. 109
☐ Teaching Transparency 3-1
☐ *Know-It Notebook* 3-1

PRACTICE AND APPLY
☐ Example 1: Average: 1–4, 9–12, 17–20, 45–51 Advanced: 1–4, 9–12, 30–33, 45–51
☐ Example 2: Average: 1–5, 9–13, 17–33, 39–40, 45–51 Advanced: 1–5, 9–13, 17–33, 41–42, 45–51
☐ Example 3: Average: 1–40, 45–51 Advanced: 1–37, 42–51

REACHING ALL LEARNERS – Differentiated Instruction for students with

Developing Knowledge	On-level Knowledge	Advanced Knowledge	English Language Development
☐ Concrete Manipulatives TE p. 109	☐ Concrete Manipulatives TE p. 109	☐ Concrete Manipulatives TE p. 109	☐ Concrete Manipulatives TE p. 109
☐ Practice A 3-1 CRB	☐ Practice B 3-1 CRB	☐ Practice C 3-1 CRB	☐ Practice A, B, or C 3-1 CRB
☐ Reteach 3-1 CRB	☐ Puzzles, Twisters & Teasers 3-1 CRB	☐ Challenge 3-1 CRB	☐ *Success for ELL* 3-1
☐ Homework Help Online Keyword: MR7 3-1	☐ Homework Help Online Keyword: MR7 3-1	☐ Homework Help Online Keyword: MR7 3-1	☐ Homework Help Online Keyword: MR7 3-1
☐ *Lesson Tutorial Video* 3-1	☐ *Lesson Tutorial Video* 3-1	☐ *Lesson Tutorial Video* 3-1	☐ *Lesson Tutorial Video* 3-1
☐ Reading Strategies 3-1 CRB	☐ Problem Solving 3-1 CRB	☐ Problem Solving 3-1 CRB	☐ Reading Strategies 3-1 CRB
☐ *Questioning Strategies* pp. 30–31	☐ Visual TE p. 109	☐ Visual TE p. 109	
☐ *IDEA Works!* 3-1			☐ *Multilingual Glossary*

ASSESSMENT
☐ Lesson Quiz, TE p. 111 and DT 3-1   ☐ State-Specific Test Prep Online Keyword: MR7 TestPrep

Copyright © Holt, Rinehart and Winston.
All rights reserved.

Holt Mathematics

Teacher's Name _____ Class _____ Date _____

Lesson Plan 3-2
Estimating Decimals pp. *112–115*   Day _____

Objective Students estimate decimal sums, differences, products, and quotients.

> **NCTM Standards:** Compute fluently and make reasonable estimates.

Pacing
☐ 45-minute Classes: 1 day   ☐ 90-minute Classes: 1/2 day   ☐ Other_____

WARM UP
☐ Warm Up TE p. 112 and Daily Transparency 3-2
☐ Problem of the Day TE p. 112 and Daily Transparency 3-2
☐ Countdown to Testing Transparency Week 5

TEACH
☐ Lesson Presentation CD-ROM 3-2
☐ Alternate Opener, Explorations Transparency 3-2, TE p. 112, and Exploration 3-2
☐ Reaching All Learners TE p. 113
☐ *Know-It Notebook* 3-2

PRACTICE AND APPLY
☐ Example 1: Average: 1, 11–12, 22, 27, 36–43 Advanced: 1, 11–12, 25, 28, 36–43
☐ Example 2: Average: 1–5, 11–16, 22–24, 36–43 Advanced: 1–5, 11–16, 25–27, 36–43
☐ Example 3: Average: 1–8, 11–19, 22–26, 36–43 Advanced: 1–8, 11–19, 27–32, 36–43
☐ Example 4: Average: 1–21, 26–30, 36–43 Advanced: 1–21, 31–43

REACHING ALL LEARNERS – Differentiated Instruction for students with

Developing Knowledge	On-level Knowledge	Advanced Knowledge	English-Language Development
☐ Curriculum Integration TE p. 113	☐ Curriculum Integration TE p. 113	☐ Curriculum Integration TE p. 113	☐ Curriculum Integration TE p. 113
☐ Practice A 3-2 CRB	☐ Practice B 3-2 CRB	☐ Practice C 3-2 CRB	☐ Practice A, B, or C 3-2 CRB
☐ Reteach 3-2 CRB	☐ Puzzles, Twisters & Teasers 3-2 CRB	☐ Challenge 3-2 CRB	☐ *Success for ELL* 3-2
☐ Homework Help Online Keyword: MR7 3-2	☐ Homework Help Online Keyword: MR7 3-2	☐ Homework Help Online Keyword: MR7 3-2	☐ Homework Help Online Keyword: MR7 3-2
☐ *Lesson Tutorial Video* 3-2	☐ *Lesson Tutorial Video* 3-2	☐ *Lesson Tutorial Video* 3-2	☐ *Lesson Tutorial Video* 3-2
☐ *Reading Strategies* 3-2 CRB	☐ *Problem Solving* 3-2 CRB	☐ *Problem Solving* 3-2 CRB	☐ *Reading Strategies* 3-2 CRB
☐ *Questioning Strategies* pp. 32–33			☐ *Lesson Vocabulary* SE p. 112
☐ *IDEA Works!* 3-2			☐ *Multilingual Glossary*

ASSESSMENT
☐ Lesson Quiz, TE p. 115 and DT 3-2   ☐ State-Specific Test Prep Online Keyword: MR7 TestPrep

Copyright © Holt, Rinehart and Winston.
All rights reserved.

Holt Mathematics

Teacher's Name _____ Class _____ Date _____

Lesson Plan 3-3
Adding and Subtracting Decimals pp. 118–121   Day _____

Objective Students add and subtract decimals.

> **NCTM Standards:** Compute fluently and make reasonable estimates.

Pacing
☐ 45-minute Classes: 1 day   ☐ 90-minute Classes: 1/2 day   ☐ Other_____

WARM UP
☐ Warm Up TE p. 118 and Daily Transparency 3-3
☐ Problem of the Day TE p. 118 and Daily Transparency 3-3
☐ Countdown to Testing Transparency Week 5

TEACH
☐ Lesson Presentation CD-ROM 3-3
☐ Alternate Opener, Explorations Transparency 3-3, TE p. 118, and Exploration 3-3
☐ Reaching All Learners TE p. 119
☐ *Technology Lab Activities* 3-3
☐ *Know-It Notebook* 3-3

PRACTICE AND APPLY
☐ Example 1: Average: 1–3, 12–13, 35, 42, 49–56 Advanced: 1–3, 12–13, 43–44, 49–56
☐ Example 2: Average: 1–7, 12–21, 30–35, 49–56 Advanced: 1–7, 12–21, 33–35, 43–45, 49–56
☐ Example 3: Average: 1–35, 42–44, 49–56 Advanced: 1–35, 45–56

REACHING ALL LEARNERS – Differentiated Instruction for students with

Developing Knowledge	On-level Knowledge	Advanced Knowledge	English Language Development
☐ Home Connection TE p. 119	☐ Home Connection TE p. 119	☐ Home Connection TE p. 119	☐ Home Connection TE p. 119
☐ Practice A 3-3 CRB	☐ Practice B 3-3 CRB	☐ Practice C 3-3 CRB	☐ Practice A, B, or C 3-3 CRB
☐ Reteach 3-3 CRB	☐ Puzzles, Twisters & Teasers 3-3 CRB	☐ Challenge 3-3 CRB	☐ *Success for ELL* 3-3
☐ Homework Help Online Keyword: MR7 3-3	☐ Homework Help Online Keyword: MR7 3-3	☐ Homework Help Online Keyword: MR7 3-3	☐ Homework Help Online Keyword: MR7 3-3
☐ *Lesson Tutorial Video* 3-3	☐ *Lesson Tutorial Video* 3-3	☐ *Lesson Tutorial Video* 3-3	☐ *Lesson Tutorial Video* 3-3
☐ Reading Strategies 3-3 CRB	☐ Problem Solving 3-3 CRB	☐ Problem Solving 3-3 CRB	☐ Reading Strategies 3-3 CRB
☐ *Questioning Strategies* pp. 34–35			
☐ *IDEA Works!* 3-3			☐ *Multilingual Glossary*

ASSESSMENT
☐ Lesson Quiz, TE p. 121 and DT 3-3   ☐ State-Specific Test Prep Online Keyword: MR7 TestPrep

Teacher's Name _____ Class _____ Date _____

Lesson Plan 3-4
Scientific Notation pp. 124–127   Day _____

Objective Students write large numbers in scientific notation.

> **NCTM Standards:** Understand numbers, ways of representing numbers, relationships among numbers, and number systems.

Pacing
- ☐ 45-minute Classes: 1 day   ☐ 90-minute Classes: 1/2 day   ☐ Other _____

WARM UP
- ☐ Warm Up TE p. 124 and Daily Transparency 3-4
- ☐ Problem of the Day TE p. 124 and Daily Transparency 3-4
- ☐ Countdown to Testing Transparency Week 5

TEACH
- ☐ Lesson Presentation CD-ROM 3-4
- ☐ Alternate Opener, Explorations Transparency 3-4, TE p. 124, and Exploration 3-4
- ☐ Reaching All Learners TE p. 125
- ☐ Teaching Transparency 3-4
- ☐ *Technology Lab Activities* 3-4
- ☐ *Know-It Notebook* 3-4

PRACTICE AND APPLY
- ☐ Example 1: Average: 1–3, 11–16, 53–62 Advanced: 1–3, 11–16, 53–62
- ☐ Example 2: Average: 1–6, 11–22, 36–44, 53–62 Advanced: 1–6, 11–22, 40–48, 53–62
- ☐ Example 3: Average: 1–9, 11–28, 30–43, 53–62 Advanced: 1–9, 11–28, 33–46, 53–62
- ☐ Example 4: Average: 1–47, 53–62 Advanced: 1–43, 48–62

REACHING ALL LEARNERS – Differentiated Instruction for students with

Developing Knowledge	On-level Knowledge	Advanced Knowledge	English Language Development
☐ Inclusion TE p. 125	☐ Inclusion TE p. 125	☐ Inclusion TE p. 125	☐ Inclusion TE p. 125
☐ Practice A 3-4 CRB	☐ Practice B 3-4 CRB	☐ Practice C 3-4 CRB	☐ Practice A, B, or C 3-4 CRB
☐ Reteach 3-4 CRB	☐ Puzzles, Twisters & Teasers 3-4 CRB	☐ Challenge 3-4 CRB	☐ *Success for ELL* 3-4
☐ Homework Help Online Keyword: MR7 3-4	☐ Homework Help Online Keyword: MR7 3-4	☐ Homework Help Online Keyword: MR7 3-4	☐ Homework Help Online Keyword: MR7 3-4
☐ *Lesson Tutorial Video* 3-4	☐ *Lesson Tutorial Video* 3-4	☐ *Lesson Tutorial Video* 3-4	☐ *Lesson Tutorial Video* 3-4
☐ Reading Strategies 3-4 CRB	☐ Problem Solving 3-4 CRB	☐ Problem Solving 3-4 CRB	☐ Reading Strategies 3-4 CRB
☐ *Questioning Strategies* pp. 36–37	☐ Communicating Math TE p. 125	☐ Communicating Math TE p. 125	☐ Lesson Vocabulary SE p. 124
☐ *IDEA Works!* 3-4			☐ *Multilingual Glossary*

ASSESSMENT
- ☐ Lesson Quiz, TE p. 127 and DT 3-4   ☐ State-Specific Test Prep Online Keyword: MR7 TestPrep

Teacher's Name _____ Class _____ Date _____

Lesson Plan 3-5
Multiplying Decimals pp. 130–133   Day _____

Objective Students multiply decimals by whole numbers and by decimals.

> **NCTM Standards:** Compute fluently and make reasonable estimates.

Pacing
- ☐ 45-minute Classes: 1 day   ☐ 90-minute Classes: 1/2 day   ☐ Other _____

WARM UP
- ☐ Warm Up TE p. 130 and Daily Transparency 3-5
- ☐ Problem of the Day TE p. 130 and Daily Transparency 3-5
- ☐ Countdown to Testing Transparency Week 6

TEACH
- ☐ Lesson Presentation CD-ROM 3-5
- ☐ Alternate Opener, Explorations Transparency 3-5, TE p. 130, and Exploration 3-5
- ☐ Reaching All Learners TE p. 131
- ☐ *Know-It Notebook* 3-5

PRACTICE AND APPLY
- ☐ Example 1: Average: 1–2, 11–12, 49, 55–65 Advanced: 1–2, 11–12, 51, 55–65
- ☐ Example 2: Average: 1–6, 11–20, 29–42, 49–50, 55–65 Advanced: 1–6, 11–20, 29–42, 51–52, 55–65
- ☐ Example 3: Average: 1–51, 55–65 Advanced: 1–48, 52–65

REACHING ALL LEARNERS – Differentiated Instruction for students with

Developing Knowledge	On-level Knowledge	Advanced Knowledge	English Language Development
☐ Cooperative Learning TE p. 131	☐ Cooperative Learning TE p. 131	☐ Cooperative Learning TE p. 131	☐ Cooperative Learning TE p. 131
☐ Practice A 3-5 CRB	☐ Practice B 3-5 CRB	☐ Practice C 3-5 CRB	☐ Practice A, B, or C 3-5 CRB
☐ Reteach 3-5 CRB	☐ Puzzles, Twisters & Teasers 3-5 CRB	☐ Challenge 3-5 CRB	☐ *Success for ELL* 3-5
☐ Homework Help Online Keyword: MR7 3-5	☐ Homework Help Online Keyword: MR7 3-5	☐ Homework Help Online Keyword: MR7 3-5	☐ Homework Help Online Keyword: MR7 3-5
☐ *Lesson Tutorial Video* 3-5	☐ *Lesson Tutorial Video* 3-5	☐ *Lesson Tutorial Video* 3-5	☐ *Lesson Tutorial Video* 3-5
☐ Reading Strategies 3-5 CRB	☐ Problem Solving 3-5 CRB	☐ Problem Solving 3-5 CRB	☐ Reading Strategies 3-5 CRB
☐ *Questioning Strategies* pp. 38–39	☐ Number Sense TE p. 131	☐ Number Sense TE p. 131	
☐ *IDEA Works!* 3-5			☐ *Multilingual Glossary*

ASSESSMENT
- ☐ Lesson Quiz, TE p. 133 and DT 3-5   ☐ State-Specific Test Prep Online Keyword: MR7 TestPrep

Holt Mathematics

Teacher's Name _____ Class _____ Date _____

Lesson Plan 3-6
Dividing Decimals by Whole Numbers pp. 134–136   Day _____

Objective Students divide decimals by whole numbers.

> **NCTM Standards:** Compute fluently and make reasonable estimates.

Pacing
- [ ] 45-minute Classes: 1 day   ☐ 90-minute Classes: 1/2 day   ☐ Other_____

WARM UP
- [ ] Warm Up TE p. 134 and Daily Transparency 3-6
- [ ] Problem of the Day TE p. 134 and Daily Transparency 3-6
- [ ] Countdown to Testing Transparency Week 6

TEACH
- [ ] Lesson Presentation CD-ROM 3-6
- [ ] Alternate Opener, Explorations Transparency 3-6, TE p. 134, and Exploration 3-6
- [ ] Reaching All Learners TE p. 135
- [ ] *Know-It Notebook* 3-6

PRACTICE AND APPLY
- [ ] Example 1: Average: 1–4, 10–13, 31–39 Advanced: 1–4, 10–13, 31–39
- [ ] Example 2: Average: 1–8, 10–17, 22–23, 31–39 Advanced: 1–8, 10–17, 24–25, 31–39
- [ ] Example 3: Average: 1–27, 31–39 Advanced: 1–25, 28–39

REACHING ALL LEARNERS – Differentiated Instruction for students with

Developing Knowledge	On-level Knowledge	Advanced Knowledge	English Language Development
☐ Visual Cues TE p. 135	☐ Visual Cues TE p. 135	☐ Visual Cues TE p. 135	☐ Visual Cues TE p. 135
☐ Practice A 3-6 CRB	☐ Practice B 3-6 CRB	☐ Practice C 3-6 CRB	☐ Practice A, B, or C 3-6 CRB
☐ Reteach 3-6 CRB	☐ Puzzles, Twisters & Teasers 3-6 CRB	☐ Challenge 3-6 CRB	☐ *Success for ELL* 3-6
☐ Homework Help Online Keyword: MR7 3-6	☐ Homework Help Online Keyword: MR7 3-6	☐ Homework Help Online Keyword: MR7 3-6	☐ Homework Help Online Keyword: MR7 3-6
☐ *Lesson Tutorial Video* 3-6	☐ *Lesson Tutorial Video* 3-6	☐ *Lesson Tutorial Video* 3-6	☐ *Lesson Tutorial Video* 3-6
☐ Reading Strategies 3-6 CRB	☐ Problem Solving 3-6 CRB	☐ Problem Solving 3-6 CRB	☐ Reading Strategies 3-6 CRB
☐ *Questioning Strategies* pp. 40–41			
☐ *IDEA Works!* 3-6			☐ *Multilingual Glossary*

ASSESSMENT
- [ ] Lesson Quiz, TE p. 136 and DT 3-6   ☐ State-Specific Test Prep Online Keyword: MR7 TestPrep

Teacher's Name _____ Class _____ Date _____

Lesson Plan 3-7
Dividing by Decimals pp. 137–140 Day _____

Objective Students divide whole numbers and decimals by decimals.

> **NCTM Standards:** Compute fluently and make reasonable estimates.

Pacing
- [ ] 45-minute Classes: 1 day
- [ ] 90-minute Classes: 1/2 day
- [ ] Other _____

WARM UP
- [ ] Warm Up TE p. 137 and Daily Transparency 3-7
- [ ] Problem of the Day TE p. 137 and Daily Transparency 3-7
- [ ] Countdown to Testing Transparency Week 6

TEACH
- [ ] Lesson Presentation CD-ROM 3-7
- [ ] Alternate Opener, Explorations Transparency 3-7, TE p. 137, and Exploration 3-7
- [ ] Reaching All Learners TE p. 138
- [ ] *Technology Lab Activities* 3-7
- [ ] *Know-It Notebook* 3-7

PRACTICE AND APPLY
- [ ] Example 1: Average: 1–6, 9–17, 21–26, 45–54 Advanced: 1–6, 9–17, 31–36, 45–54
- [ ] Example 2: Average: 1–40, 45–54 Advanced: 1–36, 41–54

REACHING ALL LEARNERS – Differentiated Instruction for students with

Developing Knowledge	On-level Knowledge	Advanced Knowledge	English Language Development
☐ Inclusion TE p. 138	☐ Kinesthetic Experience TE p. 138	☐ Kinesthetic Experience TE p. 138	☐ Kinesthetic Experience TE p. 138
☐ Practice A 3-7 CRB	☐ Practice B 3-7 CRB	☐ Practice C 3-7 CRB	☐ Practice A, B, or C 3-7 CRB
☐ Reteach 3-7 CRB	☐ Puzzles, Twisters & Teasers 3-7 CRB	☐ Challenge 3-7 CRB	☐ *Success for ELL* 3-7
☐ Homework Help Online Keyword: MR7 3-7	☐ Homework Help Online Keyword: MR7 3-7	☐ Homework Help Online Keyword: MR7 3-7	☐ Homework Help Online Keyword: MR7 3-7
☐ *Lesson Tutorial Video* 3-7	☐ *Lesson Tutorial Video* 3-7	☐ *Lesson Tutorial Video* 3-7	☐ *Lesson Tutorial Video* 3-7
☐ Reading Strategies 3-7 CRB	☐ Problem Solving 3-7 CRB	☐ Problem Solving 3-7 CRB	☐ Reading Strategies 3-7 CRB
☐ *Questioning Strategies* pp. 42–43			
☐ *IDEA Works!* 3-7			☐ *Multilingual Glossary*

ASSESSMENT
- [ ] Lesson Quiz, TE p. 140 and DT 3-7
- [ ] State-Specific Test Prep Online Keyword: MR7 TestPrep

Teacher's Name _____ Class _____ Date _____

Lesson Plan 3-8
Interpret the Quotient pp. 141–143 Day _____

Objective Students solve problems by interpreting the quotient.

> **NCTM Standards:** Understand meanings of operations and how they relate to one another; Recognize reasoning and proof as fundamental aspects of mathematics.

Pacing
☐ 45-minute Classes: 1 day   ☐ 90-minute Classes: 1/2 day   ☐ Other_____

WARM UP
☐ Warm Up TE p. 141 and Daily Transparency 3-8
☐ Problem of the Day TE p. 141 and Daily Transparency 3-8
☐ Countdown to Testing Transparency Week 6

TEACH
☐ Lesson Presentation CD-ROM 3-8
☐ Alternate Opener, Explorations Transparency 3-8, TE p. 141, and Exploration 3-8
☐ Reaching All Learners TE p. 142
☐ Teaching Transparency 3-8
☐ *Know-It Notebook* 3-8

PRACTICE AND APPLY
☐ Example 1: Average: 1, 4, 7, 13–22 Advanced: 1, 4, 8, 13–22
☐ Example 2: Average: 1–2, 4–5, 7–8, 13–22 Advanced: 1–2, 4–5, 8–9, 13–22
☐ Example 3: Average: 1–9, 13–22 Advanced: 1–6, 10–22

REACHING ALL LEARNERS – Differentiated Instruction for students with

Developing Knowledge	On-level Knowledge	Advanced Knowledge	English Language Development
☐ Critical Thinking TE p. 142	☐ Critical Thinking TE p. 142	☐ Critical Thinking TE p. 142	☐ Critical Thinking TE p. 142
☐ Practice A 3-8 CRB	☐ Practice B 3-8 CRB	☐ Practice C 3-8 CRB	☐ Practice A, B, or C 3-8 CRB
☐ Reteach 3-8 CRB	☐ Puzzles, Twisters & Teasers 3-8 CRB	☐ Challenge 3-8 CRB	☐ *Success for ELL* 3-8
☐ Homework Help Online Keyword: MR7 3-8	☐ Homework Help Online Keyword: MR7 3-8	☐ Homework Help Online Keyword: MR7 3-8	☐ Homework Help Online Keyword: MR7 3-8
☐ *Lesson Tutorial Video* 3-8	☐ *Lesson Tutorial Video* 3-8	☐ *Lesson Tutorial Video* 3-8	☐ *Lesson Tutorial Video* 3-8
☐ Reading Strategies 3-8 CRB	☐ Problem Solving 3-8 CRB	☐ Problem Solving 3-8 CRB	☐ Reading Strategies 3-8 CRB
☐ *Questioning Strategies* pp. 44–45			
☐ *IDEA Works!* 3-8			☐ *Multilingual Glossary*

ASSESSMENT
☐ Lesson Quiz, TE p. 143 and DT 3-8   ☐ State-Specific Test Prep Online Keyword: MR7 TestPrep

Copyright © Holt, Rinehart and Winston. All rights reserved.

Holt Mathematics

Teacher's Name _____ Class _____ Date _____

Lesson Plan 3-9
Solving Decimal Equations pp. 144–147 Day _____

Objective Students solve equations involving decimals.

> **NCTM Standards:** Represent and analyze mathematical situations and structures using algebraic symbols.

Pacing
- [ ] 45-minute Classes: 1 day [ ] 90-minute Classes: 1/2 day [ ] Other_____

WARM UP
- [ ] Warm Up TE p. 144 and Daily Transparency 3-9
- [ ] Problem of the Day TE p. 144 and Daily Transparency 3-9
- [ ] Countdown to Testing Transparency Week 6

TEACH
- [ ] Lesson Presentation CD-ROM 3-9
- [ ] Alternate Opener, Explorations Transparency 3-9, TE p. 144, and Exploration 3-9
- [ ] Reaching All Learners TE p. 145
- [ ] *Hands-On Lab Activities* 3-9
- [ ] *Know-It Notebook* 3-9

PRACTICE AND APPLY
- [ ] Example 1: Average: 1–6, 9–17, 20–25, 41–49 Advanced: 1–6, 9–17, 26–31, 41–49
- [ ] Example 2: Average: 1–35, 41–49 Advanced: 1–31, 36–49

REACHING ALL LEARNERS – Differentiated Instruction for students with

Developing Knowledge	On-level Knowledge	Advanced Knowledge	English Language Development
[ ] Inclusion TE p. 145	[ ] Communication TE p. 145	[ ] Communication TE p. 145	[ ] Communication TE p. 145
[ ] Practice A 3-9 CRB	[ ] Practice B 3-9 CRB	[ ] Practice C 3-9 CRB	[ ] Practice A, B, or C 3-9 CRB
[ ] Reteach 3-9 CRB	[ ] Puzzles, Twisters & Teasers 3-9 CRB	[ ] Challenge 3-9 CRB	[ ] *Success for ELL* 3-9
[ ] Homework Help Online Keyword: MR7 3-9	[ ] Homework Help Online Keyword: MR7 3-9	[ ] Homework Help Online Keyword: MR7 3-9	[ ] Homework Help Online Keyword: MR7 3-9
[ ] *Lesson Tutorial Video* 3-9	[ ] *Lesson Tutorial Video* 3-9	[ ] *Lesson Tutorial Video* 3-9	[ ] *Lesson Tutorial Video* 3-9
[ ] *Reading Strategies* 3-9 CRB	[ ] *Problem Solving* 3-9 CRB	[ ] *Problem Solving* 3-9 CRB	[ ] *Reading Strategies* 3-9 CRB
[ ] *Questioning Strategies* pp. 46–47			
[ ] *IDEA Works!* 3-9			[ ] *Multilingual Glossary*

ASSESSMENT
- [ ] Lesson Quiz, TE p. 147 and DT 3-9   [ ] State-Specific Test Prep Online Keyword: MR7 TestPrep

Teacher's Name _____ Class _____ Date _____

Lesson Plan 4-1
Divisibility pp. 164–167 Day _____

Objective Students use divisibility rules.

> **NCTM Standards:** Understand numbers, ways of representing numbers, relationships among numbers, and number systems; Develop and evaluate mathematical arguments and proofs.

Pacing
- [ ] 45-minute Classes: 1 day
- [ ] 90-minute Classes: 1/2 day
- [ ] Other_____

WARM UP
- [ ] Warm Up TE p. 164 and Daily Transparency 4-1
- [ ] Problem of the Day TE p. 164 and Daily Transparency 4-1
- [ ] Countdown to Testing Transparency Week 7

TEACH
- [ ] Lesson Presentation CD-ROM 4-1
- [ ] Alternate Opener, Explorations Transparency 4-1, TE p. 164, and Exploration 4-1
- [ ] Reaching All Learners TE p. 165
- [ ] Teaching Transparency 4-1
- [ ] *Know-It Notebook* 4-1

PRACTICE AND APPLY
- [ ] Example 1: Average: 1–4, 33–36, 48, 55–62 Advanced: 13–20, 49, 55–62
- [ ] Example 2: Average: 1–50, 55–62 Advanced: 1–46, 51–62

REACHING ALL LEARNERS – Differentiated Instruction for students with

Developing Knowledge	On-level Knowledge	Advanced Knowledge	English Language Development
☐ Kinesthetic Experience TE p. 165	☐ Kinesthetic Experience TE p. 165	☐ Kinesthetic Experience TE p. 165	☐ Kinesthetic Experience TE p. 165
☐ Practice A 4-1 CRB	☐ Practice B 4-1 CRB	☐ Practice C 4-1 CRB	☐ Practice A, B, or C 4-1 CRB
☐ Reteach 4-1 CRB	☐ Puzzles, Twisters & Teasers 4-1 CRB	☐ Challenge 4-1 CRB	☐ *Success for ELL* 4-1
☐ Homework Help Online Keyword: MR7 4-1	☐ Homework Help Online Keyword: MR7 4-1	☐ Homework Help Online Keyword: MR7 4-1	☐ Homework Help Online Keyword: MR7 4-1
☐ *Lesson Tutorial Video* 4-1	☐ *Lesson Tutorial Video* 4-1	☐ *Lesson Tutorial Video* 4-1	☐ *Lesson Tutorial Video* 4-1
☐ Reading Strategies 4-1 CRB	☐ Problem Solving 4-1 CRB	☐ Problem Solving 4-1 CRB	☐ Reading Strategies 4-1 CRB
☐ *Questioning Strategies* pp. 48–49	☐ Communicating Math TE p. 165	☐ Communicating Math TE p. 165	☐ Lesson Vocabulary SE p. 164
☐ *IDEA Works!* 4-1			☐ *Multilingual Glossary*

ASSESSMENT
- [ ] Lesson Quiz, TE p. 167 and DT 4-1
- [ ] State-Specific Test Prep Online Keyword: MR7 TestPrep

Teacher's Name _____ Class _____ Date _____

Lesson Plan 4-2
Factors and Prime Factorization pp. 169–172   Day _____

Objective Students write prime factorizations of composite numbers.

> **NCTM Standards:** Understand numbers, ways of representing numbers, relationships among numbers, and number systems.

Pacing
☐ 45-minute Classes: 1 day   ☐ 90-minute Classes: 1/2 day   ☐ Other _____

WARM UP
☐ Warm Up TE p. 169 and Daily Transparency 4-2
☐ Problem of the Day TE p. 169 and Daily Transparency 4-2
☐ Countdown to Testing Transparency Week 7

TEACH
☐ Lesson Presentation CD-ROM 4-2
☐ Alternate Opener, Explorations Transparency 4-2, TE p. 169, and Exploration 4-2
☐ Reaching All Learners TE p. 170
☐ *Know-It Notebook* 4-2

PRACTICE AND APPLY
☐ Example 1: Average: 1–4, 25–32, 34, 52–62 Advanced: 9–16, 25–28, 50, 52–62
☐ Example 2: Average: 1–45, 52–62 Advanced: 1–32, 35–42, 46–62

REACHING ALL LEARNERS – Differentiated Instruction for students with

Developing Knowledge	On-level Knowledge	Advanced Knowledge	English Language Development
☐ Critical Thinking TE p. 170	☐ Critical Thinking TE p. 170	☐ Critical Thinking TE p. 170	☐ Critical Thinking TE p. 170
☐ Practice A 4-2 CRB	☐ Practice B 4-2 CRB	☐ Practice C 4-2 CRB	☐ Practice A, B, or C 4-2 CRB
☐ Reteach 4-2 CRB	☐ Puzzles, Twisters & Teasers 4-2 CRB	☐ Challenge 4-2 CRB	☐ *Success for ELL* 4-2
☐ Homework Help Online Keyword: MR7 4-2	☐ Homework Help Online Keyword: MR7 4-2	☐ Homework Help Online Keyword: MR7 4-2	☐ Homework Help Online Keyword: MR7 4-2
☐ *Lesson Tutorial Video* 4-2	☐ *Lesson Tutorial Video* 4-2	☐ *Lesson Tutorial Video* 4-2	☐ *Lesson Tutorial Video* 4-2
☐ Reading Strategies 4-2 CRB	☐ Problem Solving 4-2 CRB	☐ Problem Solving 4-2 CRB	☐ Reading Strategies 4-2 CRB
☐ *Questioning Strategies* pp. 50–51	☐ Multiple Representations TE p. 170	☐ Multiple Representations TE p. 170	☐ Lesson Vocabulary SE p. 169
☐ *IDEA Works!* 4-2			☐ *Multilingual Glossary*

ASSESSMENT
☐ Lesson Quiz; TE p. 172 and DT 4-2   ☐ State-Specific Test Prep Online Keyword: MR7 TestPrep

Holt Mathematics

Teacher's Name _____ Class _____ Date _____

Lesson Plan 4-3
Greatest Common Factor pp. 173–176   Day _____

Objective Students find the greatest common factor (GCF) of a set of numbers.

> **NCTM Standards:** Understand numbers, ways of representing numbers, relationships among numbers, and number systems.

Pacing
- ☐ 45-minute Classes: 1 day   ☐ 90-minute Classes: 1/2 day   ☐ Other_____

WARM UP
- ☐ Warm Up TE p. 173 and Daily Transparency 4-3
- ☐ Problem of the Day TE p. 173 and Daily Transparency 4-3
- ☐ Countdown to Testing Transparency Week 7

TEACH
- ☐ Lesson Presentation CD-ROM 4-3
- ☐ Alternate Opener, Explorations Transparency 4-3, TE p. 173, and Exploration 4-3
- ☐ Reaching All Learners TE p. 174
- ☐ *Know-It Notebook* 4-3

PRACTICE AND APPLY
- ☐ Example 1: Average: 1–6, 19–27, 31–36, 43–53   Advanced: 8–16, 19–24, 31–36, 43–53
- ☐ Example 2: Average: 1–39, 43–53   Advanced: 1–36, 40–53

REACHING ALL LEARNERS – Differentiated Instruction for students with

Developing Knowledge	On-level Knowledge	Advanced Knowledge	English Language Development
☐ Inclusion TE p. 174	☐ Cooperative Learning TE p. 174	☐ Cooperative Learning TE p. 174	☐ Cooperative Learning TE p. 174
☐ Practice A 4-3 CRB	☐ Practice B 4-3 CRB	☐ Practice C 4-3 CRB	☐ Practice A, B, or C 4-3 CRB
☐ Reteach 4-3 CRB	☐ Puzzles, Twisters & Teasers 4-3 CRB	☐ Challenge 4-3 CRB	☐ *Success for ELL* 4-3
☐ Homework Help Online Keyword: MR7 4-3	☐ Homework Help Online Keyword: MR7 4-3	☐ Homework Help Online Keyword: MR7 4-3	☐ Homework Help Online Keyword: MR7 4-3
☐ *Lesson Tutorial Video* 4-3	☐ *Lesson Tutorial Video* 4-3	☐ *Lesson Tutorial Video* 4-3	☐ *Lesson Tutorial Video* 4-3
☐ Reading Strategies 4-3 CRB	☐ Problem Solving 4-3 CRB	☐ Problem Solving 4-3 CRB	☐ Reading Strategies 4-3 CRB
☐ *Questioning Strategies* pp. 52–53			☐ Lesson Vocabulary SE p. 173
☐ *IDEA Works!* 4-3			☐ *Multilingual Glossary*

ASSESSMENT
- ☐ Lesson Quiz, TE p. 176 and DT 4-3   ☐ State-Specific Test Prep Online   Keyword: MR7 TestPrep

Teacher's Name _____ Class _____ Date _____

Lesson Plan 4-4
Decimals and Fractions pp. 181–184   Day _____

Objective Students convert between decimals and fractions.

> **NCTM Standards:** Understand numbers, ways of representing numbers, relationships among numbers, and number systems; Create and use representations to organize, record, and communicate mathematical ideas; Select, apply, and translate among mathematical representations to solve problems.

Pacing
☐ 45-minute Classes: 1 day   ☐ 90-minute Classes: 1/2 day   ☐ Other _____

WARM UP
☐ Warm Up TE p. 181 and Daily Transparency 4-4
☐ Problem of the Day TE p. 181 and Daily Transparency 4-4
☐ Countdown to Testing Transparency Week 8

TEACH
☐ Lesson Presentation CD-ROM 4-4
☐ Alternate Opener, Explorations Transparency 4-4, TE p. 181, and Exploration 4-4
☐ Reaching All Learners TE p. 182
☐ Teaching Transparency 4-4
☐ *Know-It Notebook* 4-4

PRACTICE AND APPLY
☐ Example 1: Average: 1–4, 63, 67–75 Advanced: 12–19, 67–75
☐ Example 2: Average: 1–8, 38–47, 63, 67–75 Advanced: 12–27, 64, 67–75
☐ Example 3: Average: 1–42, 52–75 Advanced: 12–75

REACHING ALL LEARNERS – Differentiated Instruction for students with

Developing Knowledge	On-level Knowledge	Advanced Knowledge	English Language Development
☐ Cognitive Strategies TE p. 182	☐ Cognitive Strategies TE p. 182	☐ Cognitive Strategies TE p. 182	☐ Cognitive Strategies TE p. 182
☐ Practice A 4-4 CRB	☐ Practice B 4-4 CRB	☐ Practice C 4-4 CRB	☐ Practice A, B, or C 4-4 CRB
☐ Reteach 4-4 CRB	☐ Puzzles, Twisters & Teasers 4-4 CRB	☐ Challenge 4-4 CRB	☐ *Success for ELL* 4-4
☐ Homework Help Online Keyword: MR7 4-4	☐ Homework Help Online Keyword: MR7 4-4	☐ Homework Help Online Keyword: MR7 4-4	☐ Homework Help Online Keyword: MR7 4-4
☐ *Lesson Tutorial Video* 4-4	☐ *Lesson Tutorial Video* 4-4	☐ *Lesson Tutorial Video* 4-4	☐ *Lesson Tutorial Video* 4-4
☐ Reading Strategies 4-4 CRB	☐ Problem Solving 4-4 CRB	☐ Problem Solving 4-4 CRB	☐ Reading Strategies 4-4 CRB
☐ *Questioning Strategies* pp. 54–55	☐ Multiple Representations TE p. 182	☐ Multiple Representations TE p. 182	☐ Lesson Vocabulary SE p. 181
☐ *IDEA Works!* 4-4			☐ *Multilingual Glossary*

ASSESSMENT
☐ Lesson Quiz, TE p. 184 and DT 4-4   ☐ State-Specific Test Prep Online Keyword: MR7 TestPrep

Holt Mathematics

Teacher's Name _____ Class _____ Date _____

Lesson Plan 4-5
Equivalent Fractions pp. 186–189   Day _____

Objective Students write equivalent fractions.

> **NCTM Standards:** Understand numbers, ways of representing numbers, relationships among numbers, and number systems.

Pacing
- [ ] 45-minute Classes: 1 day   - [ ] 90-minute Classes: 1/2 day   - [ ] Other_____

WARM UP
- [ ] Warm Up TE p. 186 and Daily Transparency 4-5
- [ ] Problem of the Day TE p. 186 and Daily Transparency 4-5
- [ ] Countdown to Testing Transparency Week 8

TEACH
- [ ] Lesson Presentation CD-ROM 4-5
- [ ] Alternate Opener, Explorations Transparency 4-5, TE p. 186, and Exploration 4-5
- [ ] Reaching All Learners TE p. 187
- [ ] Teaching Transparency 4-5
- [ ] *Know-It Notebook* 4-5

PRACTICE AND APPLY
- [ ] Example 1: Average: 1–4, 37–40, 45, 51–60 Advanced: 13–20, 45, 51–60
- [ ] Example 2: Average: 1–8, 37–40, 45, 51–60 Advanced: 13–28, 51–60
- [ ] Example 3: Average: 1–47, 51–60 Advanced: 1–44, 48–60

REACHING ALL LEARNERS – Differentiated Instruction for students with

Developing Knowledge	On-level Knowledge	Advanced Knowledge	English Language Development
☐ Cooperative Learning TE p. 187	☐ Cooperative Learning TE p. 187	☐ Cooperative Learning TE p. 187	☐ Cooperative Learning TE p. 187
☐ Practice A 4-5 CRB	☐ Practice B 4-5 CRB	☐ Practice C 4-5 CRB	☐ Practice A, B, or C 4-5 CRB
☐ Reteach 4-5 CRB	☐ Puzzles, Twisters & Teasers 4-5 CRB	☐ Challenge 4-5 CRB	☐ *Success for ELL* 4-5
☐ Homework Help Online Keyword: MR7 4-5	☐ Homework Help Online Keyword: MR7 4-5	☐ Homework Help Online Keyword: MR7 4-5	☐ Homework Help Online Keyword: MR7 4-5
☐ *Lesson Tutorial Video* 4-5	☐ *Lesson Tutorial Video* 4-5	☐ *Lesson Tutorial Video* 4-5	☐ *Lesson Tutorial Video* 4-5
☐ Reading Strategies 4-5 CRB	☐ Problem Solving 4-5 CRB	☐ Problem Solving 4-5 CRB	☐ Reading Strategies 4-5 CRB
☐ Questioning Strategies pp. 56–57	☐ Technology TE p. 187	☐ Technology TE p. 187	☐ Lesson Vocabulary SE p. 186
☐ *IDEA Works!* 4-5			☐ *Multilingual Glossary*

ASSESSMENT
- [ ] Lesson Quiz, TE p. 189 and DT 4-5   - [ ] State-Specific Test Prep Online Keyword: MR7 TestPrep

Teacher's Name _____ Class _____ Date _____

Lesson Plan 4-6
Mixed Numbers and Improper Fractions pp. 192–195   Day _____

Objective Students convert between mixed numbers and improper fractions.

> **NCTM Standards:** Understand numbers, ways of representing numbers, relationships among numbers, and number systems; Recognize and apply mathematics in contexts outside of mathematics; Create and use representations to organize, record, and communicate mathematical ideas.

Pacing
☐ 45-minute Classes: 1 day   ☐ 90-minute Classes: 1/2 day   ☐ Other_____

WARM UP
☐ Warm Up TE p. 192 and Daily Transparency 4-6
☐ Problem of the Day TE p. 192 and Daily Transparency 4-6
☐ Countdown to Testing Transparency Week 9

TEACH
☐ Lesson Presentation CD-ROM 4-6
☐ Alternate Opener, Explorations Transparency 4-6, TE p. 192, and Exploration 4-6
☐ Reaching All Learners TE p. 193
☐ Teaching Transparency 4-6
☐ *Hands-On Lab Activities* 4-6
☐ *Know-It Notebook* 4-6

PRACTICE AND APPLY
☐ Example 1: Average: 1, 16–23, 35, 41, 47–57 Advanced: 6–7, 16–23, 35, 47–57
☐ Example 2: Average: 1–43, 47–57 Advanced: 1–39, 44–57

REACHING ALL LEARNERS – Differentiated Instruction for students with

Developing Knowledge	On-level Knowledge	Advanced Knowledge	English Language Development
☐ Kinesthetic Experience TE p. 193	☐ Kinesthetic Experience TE p. 193	☐ Kinesthetic Experience TE p. 193	☐ Kinesthetic Experience TE p. 193
☐ Practice A 4-6 CRB	☐ Practice B 4-6 CRB	☐ Practice C 4-6 CRB	☐ Practice A, B, or C 4-6 CRB
☐ Reteach 4-6 CRB	☐ Puzzles, Twisters & Teasers 4-6 CRB	☐ Challenge 4-6 CRB	☐ *Success for ELL* 4-6
☐ Homework Help Online Keyword: MR7 4-6	☐ Homework Help Online Keyword: MR7 4-6	☐ Homework Help Online Keyword: MR7 4-6	☐ Homework Help Online Keyword: MR7 4-6
☐ *Lesson Tutorial Video* 4-6	☐ *Lesson Tutorial Video* 4-6	☐ *Lesson Tutorial Video* 4-6	☐ *Lesson Tutorial Video* 4-6
☐ Reading Strategies 4-6 CRB	☐ Problem Solving 4-6 CRB	☐ Problem Solving 4-6 CRB	☐ Reading Strategies 4-6 CRB
☐ *Questioning Strategies* pp. 58–59			☐ *Lesson Vocabulary* SE p. 192
☐ *IDEA Works!* 4-6			☐ *Multilingual Glossary*

ASSESSMENT
☐ Lesson Quiz, TE p. 195 and DT 4-6   ☐ State-Specific Test Prep Online Keyword: MR7 TestPrep

Teacher's Name _____ Class _____ Date _____

Lesson Plan 4-7
Comparing and Ordering Fractions pp. 198–201   Day _____

Objective Students use pictures and number lines to compare and order fractions.

> **NCTM Standards:** Understand numbers, ways of representing numbers, relationships among numbers, and number systems; Select, apply, and translate among mathematical representations to solve problems.

Pacing
☐ 45-minute Classes: 1 day   ☐ 90-minute Classes: 1/2 day   ☐ Other _____

WARM UP
☐ Warm Up TE p. 198 and Daily Transparency 4-7
☐ Problem of the Day TE p. 198 and Daily Transparency 4-7
☐ Countdown to Testing Transparency Week 9

TEACH
☐ Lesson Presentation CD-ROM 4-7
☐ Alternate Opener, Explorations Transparency 4-7, TE p. 198, and Exploration 4-7
☐ Reaching All Learners TE p. 199
☐ *Hands-On Lab Activities* 4-7
☐ *Technology Lab Activities* 4-7
☐ *Know-It Notebook* 4-7

PRACTICE AND APPLY
☐ Example 1: Average: 1–4, 27–34, 43, 53–63 Advanced: 10–17, 27–30, 44, 53–63
☐ Example 2: Average: 1–5, 27–34, 43, 53–63 Advanced: 10–18, 27–30, 44, 53–63
☐ Example 3: Average: 1–49, 53–63 Advanced: 1–47, 50–63

REACHING ALL LEARNERS – Differentiated Instruction for students with

Developing Knowledge	On-level Knowledge	Advanced Knowledge	English Language Development
☐ Modeling TE p. 199	☐ Modeling TE p. 199	☐ Modeling TE p. 199	☐ Modeling TE p. 199
☐ Practice A 4-7 CRB	☐ Practice B 4-7 CRB	☐ Practice C 4-7 CRB	☐ Practice A, B, or C 4-7 CRB
☐ Reteach 4-7 CRB	☐ Puzzles, Twisters & Teasers 4-7 CRB	☐ Challenge 4-7 CRB	☐ *Success for ELL* 4-7
☐ Homework Help Online Keyword: MR7 4-7	☐ Homework Help Online Keyword: MR7 4-7	☐ Homework Help Online Keyword: MR7 4-7	☐ Homework Help Online Keyword: MR7 4-7
☐ *Lesson Tutorial Video* 4-7	☐ *Lesson Tutorial Video* 4-7	☐ *Lesson Tutorial Video* 4-7	☐ *Lesson Tutorial Video* 4-7
☐ Reading Strategies 4-7 CRB	☐ Problem Solving 4-7 CRB	☐ Problem Solving 4-7 CRB	☐ Reading Strategies 4-7 CRB
☐ *Questioning Strategies* pp. 60–61	☐ Communicating Math TE p. 199	☐ Communicating Math TE p. 199	☐ Lesson Vocabulary SE p. 198
☐ *IDEA Works!* 4-7			☐ *Multilingual Glossary*

ASSESSMENT
☐ Lesson Quiz, TE p. 201 and DT 4-7   ☐ State-Specific Test Prep Online Keyword: MR7 TestPrep

Copyright © Holt, Rinehart and Winston.
All rights reserved.

Holt Mathematics

Teacher's Name _____ Class _____ Date _____

Lesson Plan 4-8
Adding and Subtracting with Like Denominators pp. 202–205   Day _____

Objective Students add and subtract fractions with like denominators.

> **NCTM Standards:** Compute fluently and make reasonable estimates; Recognize and apply mathematics in contexts outside of mathematics; Select and use various types of reasoning and methods of proof.

Pacing
☐ 45-minute Classes: 1 day   ☐ 90-minute Classes: 1/2 day   ☐ Other_____

WARM UP
☐ Warm Up TE p. 202 and Daily Transparency 4-8
☐ Problem of the Day TE p. 202 and Daily Transparency 4-8
☐ Countdown to Testing Transparency Week 9

TEACH
☐ Lesson Presentation CD-ROM 4-8
☐ Alternate Opener, Explorations Transparency 4-8, TE p. 202, and Exploration 4-8
☐ Reaching All Learners TE p. 203
☐ *Know-It Notebook* 4-8

PRACTICE AND APPLY
☐ Example 1: Average: 1, 28, 38, 44–53 Advanced: 10, 28, 38, 44–53
☐ Example 2: Average: 1–5, 19–28, 36–38, 44–53 Advanced: 10–14, 19–28, 40–42, 44–53
☐ Example 3: Average: 1–35, 39–40, 44–53 Advanced: 1–35, 41–53

REACHING ALL LEARNERS – Differentiated Instruction for students with

Developing Knowledge	On-level Knowledge	Advanced Knowledge	English Language Development
☐ Critical Thinking TE p. 203	☐ Critical Thinking TE p. 203	☐ Critical Thinking TE p. 203	☐ Critical Thinking TE p. 203
☐ Practice A 4-8 CRB	☐ Practice B 4-8 CRB	☐ Practice C 4-8 CRB	☐ Practice A, B, or C 4-8 CRB
☐ Reteach 4-8 CRB	☐ Puzzles, Twisters & Teasers 4-8 CRB	☐ Challenge 4-8 CRB	☐ *Success for ELL* 4-8
☐ Homework Help Online Keyword: MR7 4-8	☐ Homework Help Online Keyword: MR7 4-8	☐ Homework Help Online Keyword: MR7 4-8	☐ Homework Help Online Keyword: MR7 4-8
☐ *Lesson Tutorial Video* 4-8	☐ *Lesson Tutorial Video* 4-8	☐ *Lesson Tutorial Video* 4-8	☐ *Lesson Tutorial Video* 4-8
☐ Reading Strategies 4-8 CRB	☐ Problem Solving 4-8 CRB	☐ Problem Solving 4-8 CRB	☐ Reading Strategies 4-8 CRB
☐ Questioning Strategies pp. 62–63			
☐ *IDEA Works!* 4-8			☐ *Multilingual Glossary*

ASSESSMENT
☐ Lesson Quiz, TE p. 205 and DT 4-8   ☐ State-Specific Test Prep Online Keyword: MR7 TestPrep

Teacher's Name _____ Class _____ Date _____

Lesson Plan 4-9
Estimating Fraction Sums and Differences pp. 206–209 Day _____

Objective Students estimate sums and differences of fractions and mixed numbers.

> **NCTM Standards:** Compute fluently and make reasonable estimates; Select and use various types of reasoning and methods of proof.

Pacing
☐ 45-minute Classes: 1 day   ☐ 90-minute Classes: 1/2 day   ☐ Other _____

WARM UP
☐ Warm Up TE p. 206 and Daily Transparency 4-9
☐ Problem of the Day TE p. 206 and Daily Transparency 4-9
☐ Countdown to Testing Transparency Week 9

TEACH
☐ Lesson Presentation CD-ROM 4-9
☐ Alternate Opener, Explorations Transparency 4-9, TE p. 206, and Exploration 4-9
☐ Reaching All Learners TE p. 207
☐ Teaching Transparency 4-9
☐ *Know-It Notebook* 4-9

PRACTICE AND APPLY
☐ Example 1: Average: 1–4, 37–43 Advanced: 7–10, 37–43
☐ Example 2: Average: 1–21, 27–33, 37–43 Advanced: 7–30, 34–43

REACHING ALL LEARNERS – Differentiated Instruction for students with

Developing Knowledge	On-level Knowledge	Advanced Knowledge	English Language Development
☐ Curriculum Integration TE p. 207	☐ Curriculum Integration TE p. 207	☐ Curriculum Integration TE p. 207	☐ Curriculum Integration TE p. 207
☐ Practice A 4-9 CRB	☐ Practice B 4-9 CRB	☐ Practice C 4-9 CRB	☐ Practice A, B, or C 4-9 CRB
☐ Reteach 4-9 CRB	☐ Puzzles, Twisters & Teasers 4-9 CRB	☐ Challenge 4-9 CRB	☐ *Success for ELL* 4-9
☐ Homework Help Online Keyword: MR7 4-9	☐ Homework Help Online Keyword: MR7 4-9	☐ Homework Help Online Keyword: MR7 4-9	☐ Homework Help Online Keyword: MR7 4-9
☐ *Lesson Tutorial Video* 4-9	☐ *Lesson Tutorial Video* 4-9	☐ *Lesson Tutorial Video* 4-9	☐ *Lesson Tutorial Video* 4-9
☐ Reading Strategies 4-9 CRB	☐ Problem Solving 4-9 CRB	☐ Problem Solving 4-9 CRB	☐ Reading Strategies 4-9 CRB
☐ *Questioning Strategies* pp. 64–65			
☐ *IDEA Works!* 4-9			☐ *Multilingual Glossary*

ASSESSMENT
☐ Lesson Quiz, TE p. 209 and DT 4-9   ☐ State-Specific Test Prep Online Keyword: MR7 TestPrep

Teacher's Name _____ Class _____ Date _____

Lesson Plan 5-1
Least Common Multiple pp. 228–231   Day _____

Objective Students find the least common multiple (LCM) of a group of numbers.

> **NCTM Standards:** Understand numbers, ways of representing numbers, relationships among numbers, and number systems.

Pacing
☐ 45-minute Classes: 1 day   ☐ 90-minute Classes: 1/2 day   ☐ Other_____

WARM UP
☐ Warm Up TE p. 228 and Daily Transparency 5-1
☐ Problem of the Day TE p. 228 and Daily Transparency 5-1
☐ Countdown to Testing Transparency Week 10

TEACH
☐ Lesson Presentation CD-ROM 5-1
☐ Alternate Opener, Explorations Transparency 5-1, TE p. 228, and Exploration 5-1
☐ Reaching All Learners TE p. 229
☐ *Technology Lab Activities* 5-1
☐ *Know-It Notebook* 5-1

PRACTICE AND APPLY
☐ Example 1: Average: 1, 41–52 Advanced: 14, 41–52
☐ Example 2: Average: 1–38, 41–52 Advanced: 1–36, 38–52

REACHING ALL LEARNERS – Differentiated Instruction for students with

Developing Knowledge	On-level Knowledge	Advanced Knowledge	English Language Development
☐ Concrete Manipulatives TE p. 229	☐ Concrete Manipulatives TE p. 229	☐ Concrete Manipulatives TE p. 229	☐ Concrete Manipulatives TE p. 229
☐ Practice A 5-1 CRB	☐ Practice B 5-1 CRB	☐ Practice C 5-1 CRB	☐ Practice A, B, or C 5-1 CRB
☐ Reteach 5-1 CRB	☐ Puzzles, Twisters & Teasers 5-1 CRB	☐ Challenge 5-1 CRB	☐ *Success for ELL* 5-1
☐ Homework Help Online Keyword: MR7 5-1	☐ Homework Help Online Keyword: MR7 5-1	☐ Homework Help Online Keyword: MR7 5-1	☐ Homework Help Online Keyword: MR7 5-1
☐ *Lesson Tutorial Video* 5-1	☐ *Lesson Tutorial Video* 5-1	☐ *Lesson Tutorial Video* 5-1	☐ *Lesson Tutorial Video* 5-1
☐ Reading Strategies 5-1 CRB	☐ Problem Solving 5-1 CRB	☐ Problem Solving 5-1 CRB	☐ Reading Strategies 5-1 CRB
☐ *Questioning Strategies* pp. 66–67	☐ Multiple Representations TE p. 229	☐ Multiple Representations TE p. 229	☐ Lesson Vocabulary SE p. 228
☐ *IDEA Works!* 5-1			☐ *Multilingual Glossary*

ASSESSMENT
☐ Lesson Quiz, TE p. 231 and DT 5-1   ☐ State-Specific Test Prep Online Keyword: MR7 TestPrep

Teacher's Name _____ Class _____ Date _____

Lesson Plan 5-2
Adding and Subtracting with Unlike Denominators pp. 234–237   Day _____

Objective Students add and subtract fractions with unlike denominators.

> **NCTM Standards:** Understand meanings of operations and how they relate to one another; Select and use various types of reasoning and methods of proof.

Pacing
☐ 45-minute Classes: 1 day   ☐ 90-minute Classes: 1/2 day   ☐ Other _____

WARM UP
☐ Warm Up TE p. 234 and Daily Transparency 5-2
☐ Problem of the Day TE p. 234 and Daily Transparency 5-2
☐ Countdown to Testing Transparency Week 10

TEACH
☐ Lesson Presentation CD-ROM 5-2
☐ Alternate Opener, Explorations Transparency 5-2, TE p. 234, and Exploration 5-2
☐ Reaching All Learners TE p. 235
☐ Teaching Transparency 5-2
☐ *Technology Lab Activities* 5-2
☐ *Know-It Notebook* 5-2

PRACTICE AND APPLY
☐ Example 1: Average: 1, 48–57 Advanced: 6–7, 48–57
☐ Example 2: Average: 1–43, 48–57 Advanced: 1–39, 44–57

REACHING ALL LEARNERS – Differentiated Instruction for students with

Developing Knowledge	On-level Knowledge	Advanced Knowledge	English Language Development
☐ Modeling TE p. 235	☐ Modeling TE p. 235	☐ Modeling TE p. 235	☐ Modeling TE p. 235
☐ Practice A 5-2 CRB	☐ Practice B 5-2 CRB	☐ Practice C 5-2 CRB	☐ Practice A, B, or C 5-2 CRB
☐ Reteach 5-2 CRB	☐ Puzzles, Twisters & Teasers 5-2 CRB	☐ Challenge 5-2 CRB	☐ *Success for ELL* 5-2
☐ Homework Help Online Keyword: MR7 5-2	☐ Homework Help Online Keyword: MR7 5-2	☐ Homework Help Online Keyword: MR7 5-2	☐ Homework Help Online Keyword: MR7 5-2
☐ *Lesson Tutorial Video* 5-2	☐ *Lesson Tutorial Video* 5-2	☐ *Lesson Tutorial Video* 5-2	☐ *Lesson Tutorial Video* 5-2
☐ Reading Strategies 5-2 CRB	☐ Problem Solving 5-2 CRB	☐ Problem Solving 5-2 CRB	☐ Reading Strategies 5-2 CRB
☐ *Questioning Strategies* pp. 68–69	☐ Cognitive Strategies TE p. 235	☐ Cognitive Strategies TE p. 235	☐ Lesson Vocabulary SE p. 234
☐ *IDEA Works!* 5-2			☐ *Multilingual Glossary*

ASSESSMENT
☐ Lesson Quiz, TE p. 237 and DT 5-2   ☐ State-Specific Test Prep Online Keyword: MR7 TestPrep

Teacher's Name _____ Class _____ Date _____

Lesson Plan 5-3
Adding and Subtracting Mixed Numbers pp. 238–241   Day _____

Objective Students add and subtract mixed numbers with unlike denominators.

> **NCTM Standards:** Understand meanings of operations and how they relate to one another.

Pacing
☐ 45-minute Classes: 1 day   ☐ 90-minute Classes: 1/2 day   ☐ Other_____

WARM UP
☐ Warm Up TE p. 238 and Daily Transparency 5-3
☐ Problem of the Day TE p. 238 and Daily Transparency 5-3
☐ Countdown to Testing Transparency Week 10

TEACH
☐ Lesson Presentation CD-ROM 5-3
☐ Alternate Opener, Explorations Transparency 5-3, TE p. 238, and Exploration 5-3
☐ Reaching All Learners TE p. 239
☐ *Technology Lab Activities* 5-3
☐ *Know-It Notebook* 5-3

PRACTICE AND APPLY
☐ Example 1: Average: 1–4, 15–22, 50–59 Advanced: 6–13, 24–27, 50–59
☐ Example 2: Average: 1–44, 50–59 Advanced: 1–42, 45–59

REACHING ALL LEARNERS – Differentiated Instruction for students with

Developing Knowledge	On-level Knowledge	Advanced Knowledge	English Language Development
☐ Kinesthetic Experience TE p. 239	☐ Kinesthetic Experience TE p. 239	☐ Kinesthetic Experience TE p. 239	☐ Kinesthetic Experience TE p. 239
☐ Practice A 5-3 CRB	☐ Practice B 5-3 CRB	☐ Practice C 5-3 CRB	☐ Practice A, B, or C 5-3 CRB
☐ Reteach 5-3 CRB	☐ Puzzles, Twisters & Teasers 5-3 CRB	☐ Challenge 5-3 CRB	☐ *Success for ELL* 5-3
☐ Homework Help Online Keyword: MR7 5-3	☐ Homework Help Online Keyword: MR7 5-3	☐ Homework Help Online Keyword: MR7 5-3	☐ Homework Help Online Keyword: MR7 5-3
☐ *Lesson Tutorial Video* 5-3	☐ *Lesson Tutorial Video* 5-3	☐ *Lesson Tutorial Video* 5-3	☐ *Lesson Tutorial Video* 5-3
☐ Reading Strategies 5-3 CRB	☐ Problem Solving 5-3 CRB	☐ Problem Solving 5-3 CRB	☐ Reading Strategies 5-3 CRB
☐ *Questioning Strategies* pp. 70–71			
☐ *IDEA Works!* 5-3			☐ *Multilingual Glossary*

ASSESSMENT
☐ Lesson Quiz, TE p. 241 and DT 5-3   ☐ State-Specific Test Prep Online Keyword: MR7 TestPrep

Teacher's Name _____ Class _____ Date _____

Lesson Plan 5-4
Regrouping to Subtract Mixed Numbers pp. 244–247 Day _____

Objective Students regroup mixed numbers to subtract.

> **NCTM Standards:** Understand meanings of operations and how they relate to one another; Recognize and use connections among mathematical ideas; Understand how mathematical ideas interconnect and build on one another to produce a coherent whole.

Pacing
- ☐ 45-minute Classes: 1 day ☐ 90-minute Classes: 1/2 day ☐ Other _____

WARM UP
- ☐ Warm Up TE p. 244 and Daily Transparency 5-4
- ☐ Problem of the Day TE p. 244 and Daily Transparency 5-4
- ☐ Countdown to Testing Transparency Week 10

TEACH
- ☐ Lesson Presentation CD-ROM 5-4
- ☐ Alternate Opener, Explorations Transparency 5-4, TE p. 244, and Exploration 5-4
- ☐ Reaching All Learners TE p. 245
- ☐ Teaching Transparency 5-4
- ☐ *Know-It Notebook* 5-4

PRACTICE AND APPLY
- ☐ Example 1: Average: 1–4, 16–23, 47–55 Advanced: 6–13, 26–29, 47–55
- ☐ Example 2: Average: 1–43, 47–55 Advanced: 1–40, 44–55

REACHING ALL LEARNERS – Differentiated Instruction for students with

Developing Knowledge	On-level Knowledge	Advanced Knowledge	English Language Development
☐ Critical Thinking TE p. 245	☐ Critical Thinking TE p. 245	☐ Critical Thinking TE p. 245	☐ Critical Thinking TE p. 245
☐ Practice A 5-4 CRB	☐ Practice B 5-4 CRB	☐ Practice C 5-4 CRB	☐ Practice A, B, or C 5-4 CRB
☐ Reteach 5-4 CRB	☐ Puzzles, Twisters & Teasers 5-4 CRB	☐ Challenge 5-4 CRB	☐ *Success for ELL* 5-4
☐ Homework Help Online Keyword: MR7 5-4	☐ Homework Help Online Keyword: MR7 5-4	☐ Homework Help Online Keyword: MR7 5-4	☐ Homework Help Online Keyword: MR7 5-4
☐ *Lesson Tutorial Video* 5-4	☐ *Lesson Tutorial Video* 5-4	☐ *Lesson Tutorial Video* 5-4	☐ *Lesson Tutorial Video* 5-4
☐ Reading Strategies 5-4 CRB	☐ Problem Solving 5-4 CRB	☐ Problem Solving 5-4 CRB	☐ Reading Strategies 5-4 CRB
☐ *Questioning Strategies* pp. 72–73			
☐ *IDEA Works!* 5-4			☐ *Multilingual Glossary*

ASSESSMENT
- ☐ Lesson Quiz, TE p. 247 and DT 5-4 ☐ State-Specific Test Prep Online Keyword: MR7 TestPrep

Teacher's Name _____ Class _____ Date _____

Lesson Plan 5-5
Solving Fraction Equations: Addition and Subtraction pp. 248–251   Day _____

Objective Students solve equations by adding and subtracting fractions.

> **NCTM Standards:** Understand meanings of operations and how they relate to one another; Represent and analyze mathematical situations and structures using algebraic symbols.

Pacing
☐ 45-minute Classes: 1 day   ☐ 90-minute Classes: 1/2 day   ☐ Other_____

WARM UP
☐ Warm Up TE p. 248 and Daily Transparency 5-5
☐ Problem of the Day TE p. 248 and Daily Transparency 5-5
☐ Countdown to Testing Transparency Week 10

TEACH
☐ Lesson Presentation CD-ROM 5-5
☐ Alternate Opener, Explorations Transparency 5-5, TE p. 248, and Exploration 5-5
☐ Reaching All Learners TE p. 249
☐ *Know-It Notebook* 5-5

PRACTICE AND APPLY
☐ Example 1: Average: 1–6, 16–21, 37–45 Advanced: 8–13, 26–31, 37–45
☐ Example 2: Average: 1–33, 37–45 Advanced: 1–31, 34–45

REACHING ALL LEARNERS – Differentiated Instruction for students with

Developing Knowledge	On-level Knowledge	Advanced Knowledge	English Language Development
☐ Diversity TE p. 249	☐ Diversity TE p. 249	☐ Diversity TE p. 249	☐ Diversity TE p. 249
☐ Practice A 5-5 CRB	☐ Practice B 5-5 CRB	☐ Practice C 5-5 CRB	☐ Practice A, B, or C 5-5 CRB
☐ Reteach 5-5 CRB	☐ Puzzles, Twisters & Teasers 5-5 CRB	☐ Challenge 5-5 CRB	☐ *Success for ELL* 5-5
☐ Homework Help Online Keyword: MR7 5-5	☐ Homework Help Online Keyword: MR7 5-5	☐ Homework Help Online Keyword: MR7 5-5	☐ Homework Help Online Keyword: MR7 5-5
☐ *Lesson Tutorial Video* 5-5	☐ *Lesson Tutorial Video* 5-5	☐ *Lesson Tutorial Video* 5-5	☐ *Lesson Tutorial Video* 5-5
☐ Reading Strategies 5-5 CRB	☐ Problem Solving 5-5 CRB	☐ Problem Solving 5-5 CRB	☐ Reading Strategies 5-5 CRB
☐ *Questioning Strategies* pp. 74–75			
☐ *IDEA Works!* 5-5			☐ *Multilingual Glossary*

ASSESSMENT
☐ Lesson Quiz, TE p. 251 and DT 5-5   ☐ State-Specific Test Prep Online Keyword: MR7 TestPrep

Teacher's Name _____ Class _____ Date _____

Lesson Plan 5-6
Multiplying Fractions by Whole Numbers pp. 254–257   Day _____

Objective Students multiply fractions by whole numbers.

NCTM Standards: Compute fluently and make reasonable estimates.

Pacing
☐ 45-minute Classes: 1 day   ☐ 90-minute Classes: 1/2 day   ☐ Other_____

WARM UP
☐ Warm Up TE p. 254 and Daily Transparency 5-6
☐ Problem of the Day TE p. 254 and Daily Transparency 5-6
☐ Countdown to Testing Transparency Week 11

TEACH
☐ Lesson Presentation CD-ROM 5-6
☐ Alternate Opener, Explorations Transparency 5-6, TE p. 254, and Exploration 5-6
☐ Reaching All Learners TE p. 255
☐ Teaching Transparency 5-6
☐ *Know-It Notebook* 5-6

PRACTICE AND APPLY
☐ Example 1: Average: 1–8, 55–63 Advanced: 14–21, 55–63
☐ Example 2: Average: 1–12, 31–34, 55–63 Advanced: 14–29, 55–63
☐ Example 3: Average: 1–51, 55–63 Advanced: 1–48, 52–63

REACHING ALL LEARNERS – Differentiated Instruction for students with

Developing Knowledge	On-level Knowledge	Advanced Knowledge	English Language Development
☐ Inclusion TE p. 255	☐ Cooperative Learning TE p. 255	☐ Cooperative Learning TE p. 255	☐ Cooperative Learning TE p. 255
☐ Practice A 5-6 CRB	☐ Practice B 5-6 CRB	☐ Practice C 5-6 CRB	☐ Practice A, B, or C 5-6 CRB
☐ Reteach 5-6 CRB	☐ Puzzles, Twisters & Teasers 5-6 CRB	☐ Challenge 5-6 CRB	☐ *Success for ELL* 5-6
☐ Homework Help Online Keyword: MR7 5-6	☐ Homework Help Online Keyword: MR7 5-6	☐ Homework Help Online Keyword: MR7 5-6	☐ Homework Help Online Keyword: MR7 5-6
☐ *Lesson Tutorial Video* 5-6	☐ *Lesson Tutorial Video* 5-6	☐ *Lesson Tutorial Video* 5-6	☐ *Lesson Tutorial Video* 5-6
☐ Reading Strategies 5-6 CRB	☐ Problem Solving 5-6 CRB	☐ Problem Solving 5-6 CRB	☐ Reading Strategies 5-6 CRB
☐ *Questioning Strategies* pp. 76–77			
☐ *IDEA Works!* 5-6			☐ *Multilingual Glossary*

ASSESSMENT
☐ Lesson Quiz, TE p. 257 and DT 5-6   ☐ State-Specific Test Prep Online Keyword: MR7 TestPrep

Teacher's Name _____ Class _____ Date _____

Lesson Plan 5-7
Multiplying Fractions pp. 260–263   Day _____

Objective Students multiply fractions.

> **NCTM Standards:** Understand meanings of operations and how they relate to one another; Select and use various types of reasoning and methods of proof.

Pacing
☐ 45-minute Classes: 1 day   ☐ 90-minute Classes: 1/2 day   ☐ Other _____

WARM UP
☐ Warm Up TE p. 260 and Daily Transparency 5-7
☐ Problem of the Day TE p. 260 and Daily Transparency 5-7
☐ Countdown to Testing Transparency Week 11

TEACH
☐ Lesson Presentation CD-ROM 5-7
☐ Alternate Opener, Explorations Transparency 5-7, TE p. 260, and Exploration 5-7
☐ Reaching All Learners TE p. 261
☐ Teaching Transparency 5-7
☐ *Technology Lab Activities* 5-7
☐ *Know-It Notebook* 5-7

PRACTICE AND APPLY
☐ Example 1: Average: 1–4, 25–28, 49–58 Advanced: 9–16, 49–58
☐ Example 2: Average: 1–44, 49–58 Advanced: 1–40, 45–58

REACHING ALL LEARNERS – Differentiated Instruction for students with

Developing Knowledge	On-level Knowledge	Advanced Knowledge	English Language Development
☐ Kinesthetic Experience TE p. 261	☐ Kinesthetic Experience TE p. 261	☐ Kinesthetic Experience TE p. 261	☐ Kinesthetic Experience TE p. 261
☐ Practice A 5-7 CRB	☐ Practice B 5-7 CRB	☐ Practice C 5-7 CRB	☐ Practice A, B, or C 5-7 CRB
☐ Reteach 5-7 CRB	☐ Puzzles, Twisters & Teasers 5-7 CRB	☐ Challenge 5-7 CRB	☐ *Success for ELL* 5-7
☐ Homework Help Online Keyword: MR7 5-7	☐ Homework Help Online Keyword: MR7 5-7	☐ Homework Help Online Keyword: MR7 5-7	☐ Homework Help Online Keyword: MR7 5-7
☐ *Lesson Tutorial Video* 5-7	☐ *Lesson Tutorial Video* 5-7	☐ *Lesson Tutorial Video* 5-7	☐ *Lesson Tutorial Video* 5-7
☐ Reading Strategies 5-7 CRB	☐ Problem Solving 5-7 CRB	☐ Problem Solving 5-7 CRB	☐ Reading Strategies 5-7 CRB
☐ *Questioning Strategies* pp. 78–79	☐ Multiple Representations TE p. 261	☐ Multiple Representations TE p. 261	
☐ *IDEA Works!* 5-7			☐ *Multilingual Glossary*

ASSESSMENT
☐ Lesson Quiz, TE p. 263 and DT 5-7   ☐ State-Specific Test Prep Online Keyword: MR7 TestPrep

Copyright © Holt, Rinehart and Winston.
All rights reserved.

Holt Mathematics

Teacher's Name _____ Class _____ Date _____

Lesson Plan 5-8
Multiplying Mixed Numbers pp. 264–267   Day _____

Objective Students multiply mixed numbers.

> **NCTM Standards:** Understand meanings of operations and how they relate to one another.

Pacing
☐ 45-minute Classes: 1 day   ☐ 90-minute Classes: 1/2 day   ☐ Other_____

WARM UP
☐ Warm Up TE p. 264 and Daily Transparency 5-8
☐ Problem of the Day TE p. 264 and Daily Transparency 5-8
☐ Countdown to Testing Transparency Week 11

TEACH
☐ Lesson Presentation CD-ROM 5-8
☐ Alternate Opener, Explorations Transparency 5-8, TE p. 264, and Exploration 5-8
☐ Reaching All Learners TE p. 265
☐ *Know-It Notebook* 5-8

PRACTICE AND APPLY
☐ Example 1: Average: 1–6, 29–33, 59–68 Advanced: 13–20, 29–31, 59–68
☐ Example 2: Average: 1–55, 59–68 Advanced: 1–52, 56–68

REACHING ALL LEARNERS – Differentiated Instruction for students with

Developing Knowledge	On-level Knowledge	Advanced Knowledge	English Language Development
☐ Critical Thinking TE p. 265	☐ Critical Thinking TE p. 265	☐ Critical Thinking TE p. 265	☐ Critical Thinking TE p. 265
☐ Practice A 5-8 CRB	☐ Practice B 5-8 CRB	☐ Practice C 5-8 CRB	☐ Practice A, B, or C 5-8 CRB
☐ Reteach 5-8 CRB	☐ Puzzles, Twisters & Teasers 5-8 CRB	☐ Challenge 5-8 CRB	☐ *Success for ELL* 5-8
☐ Homework Help Online Keyword: MR7 5-8	☐ Homework Help Online Keyword: MR7 5-8	☐ Homework Help Online Keyword: MR7 5-8	☐ Homework Help Online Keyword: MR7 5-8
☐ *Lesson Tutorial Video* 5-8	☐ *Lesson Tutorial Video* 5-8	☐ *Lesson Tutorial Video* 5-8	☐ *Lesson Tutorial Video* 5-8
☐ Reading Strategies 5-8 CRB	☐ Problem Solving 5-8 CRB	☐ Problem Solving 5-8 CRB	☐ Reading Strategies 5-8 CRB
☐ *Questioning Strategies* pp. 80–81	☐ Modeling TE p. 265	☐ Modeling TE p. 265	
☐ *IDEA Works!* 5-8			☐ *Multilingual Glossary*

ASSESSMENT
☐ Lesson Quiz, TE p. 267 and DT 5-8   ☐ State-Specific Test Prep Online Keyword: MR7 TestPrep

Teacher's Name _____ Class _____ Date _____

Lesson Plan 5-9
Dividing Fractions and Mixed Numbers pp. 270–273   Day _____

Objective Students divide fractions and mixed numbers.

> **NCTM Standards:** Understand meanings of operations and how they relate to one another; Develop and evaluate mathematical arguments and proofs.

Pacing
☐ 45-minute Classes: 1 day   ☐ 90-minute Classes: 1/2 day   ☐ Other _____

WARM UP
☐ Warm Up TE p. 270 and Daily Transparency 5-9
☐ Problem of the Day TE p. 270 and Daily Transparency 5-9
☐ Countdown to Testing Transparency Week 11

TEACH
☐ Lesson Presentation CD-ROM 5-9
☐ Alternate Opener, Explorations Transparency 5-9, TE p. 270, and Exploration 5-9
☐ Reaching All Learners TE p. 271
☐ *Know-It Notebook* 5-9

PRACTICE AND APPLY
☐ Example 1: Average: 1–5, 66–75 Advanced: 14–23, 66–75
☐ Example 2: Average: 1–56, 60–61, 66–75 Advanced: 1–53, 57–59, 62–75

REACHING ALL LEARNERS – Differentiated Instruction for students with

Developing Knowledge	On-level Knowledge	Advanced Knowledge	English Language Development
☐ Critical Thinking TE p. 271	☐ Critical Thinking TE p. 271	☐ Critical Thinking TE p. 271	☐ Critical Thinking TE p. 271
☐ Practice A 5-9 CRB	☐ Practice B 5-9 CRB	☐ Practice C 5-9 CRB	☐ Practice A, B, or C 5-9 CRB
☐ Reteach 5-9 CRB	☐ Puzzles, Twisters & Teasers 5-9 CRB	☐ Challenge 5-9 CRB	☐ *Success for ELL* 5-9
☐ Homework Help Online Keyword: MR7 5-9	☐ Homework Help Online Keyword: MR7 5-9	☐ Homework Help Online Keyword: MR7 5-9	☐ Homework Help Online Keyword: MR7 5-9
☐ *Lesson Tutorial Video* 5-9	☐ *Lesson Tutorial Video* 5-9	☐ *Lesson Tutorial Video* 5-9	☐ *Lesson Tutorial Video* 5-9
☐ Reading Strategies 5-9 CRB	☐ Problem Solving 5-9 CRB	☐ Problem Solving 5-9 CRB	☐ Reading Strategies 5-9 CRB
☐ *Questioning Strategies* pp. 82–83			☐ Lesson Vocabulary SE p. 270
☐ *IDEA Works!* 5-9			☐ *Multilingual Glossary*

ASSESSMENT
☐ Lesson Quiz, TE p. 273 and DT 5-9   ☐ State-Specific Test Prep Online Keyword: MR7 TestPrep

Teacher's Name _____ Class _____ Date _____

Lesson Plan 5-10
Solving Fraction Equations: Multiplication and Division pp. 274–277 Day _____

Objective Students solve equations by multiplying and dividing fractions.

> **NCTM Standards:** Understand meanings of operations and how they relate to one another; Represent and analyze mathematical situations and structures using algebraic symbols.

Pacing
☐ 45-minute Classes: 1 day   ☐ 90-minute Classes: 1/2 day   ☐ Other_____

WARM UP
☐ Warm Up TE p. 274 and Daily Transparency 5-10
☐ Problem of the Day TE p. 274 and Daily Transparency 5-10
☐ Countdown to Testing Transparency Week 11

TEACH
☐ Lesson Presentation CD-ROM 5-10
☐ Alternate Opener, Explorations Transparency 5-10, TE p. 274, and Exploration 5-10
☐ Reaching All Learners TE p. 275
☐ *Know-It Notebook* 5-10

PRACTICE AND APPLY
☐ Example 1: Average: 1–4, 16–21, 41–51 Advanced: 6–13, 22–23, 41–51
☐ Example 2: Average: 1–35, 41–51 Advanced: 1–31, 36–51

REACHING ALL LEARNERS – Differentiated Instruction for students with

Developing Knowledge	On-level Knowledge	Advanced Knowledge	English Language Development
☐ Cooperative Learning TE p. 275	☐ Cooperative Learning TE p. 275	☐ Cooperative Learning TE p. 275	☐ Cooperative Learning TE p. 275
☐ Practice A 5-10 CRB	☐ Practice B 5-10 CRB	☐ Practice C 5-10 CRB	☐ Practice A, B, or C 5-10 CRB
☐ Reteach 5-10 CRB	☐ Puzzles, Twisters & Teasers 5-10 CRB	☐ Challenge 5-10 CRB	☐ *Success for ELL* 5-10
☐ Homework Help Online Keyword: MR7 5-10	☐ Homework Help Online Keyword: MR7 5-10	☐ Homework Help Online Keyword: MR7 5-10	☐ Homework Help Online Keyword: MR7 5-10
☐ *Lesson Tutorial Video* 5-10	☐ *Lesson Tutorial Video* 5-10	☐ *Lesson Tutorial Video* 5-10	☐ *Lesson Tutorial Video* 5-10
☐ Reading Strategies 5-10 CRB	☐ Problem Solving 5-10 CRB	☐ Problem Solving 5-10 CRB	☐ Reading Strategies 5-10 CRB
☐ *Questioning Strategies* pp. 84–85			
☐ *IDEA Works!* 5-10			☐ *Multilingual Glossary*

ASSESSMENT
☐ Lesson Quiz, TE p. 277 and DT 5-10   ☐ State-Specific Test Prep Online Keyword: MR7 TestPrep

Teacher's Name _____ Class _____ Date _____

Lesson Plan 6-1
Make a Table pp. 294–296   Day _____

Objective Students use tables to record and organize data.

> **NCTM Standards:** Formulate questions that can be addressed with data and collect, organize, and display relevant data to answer them.

Pacing
☐ 45-minute Classes: 1 day   ☐ 90-minute Classes: 1/2 day   ☐ Other _____

WARM UP
☐ Warm Up TE p. 294 and Daily Transparency 6-1
☐ Problem of the Day TE p. 294 and Daily Transparency 6-1
☐ Countdown to Testing Transparency Week 12

TEACH
☐ Lesson Presentation CD-ROM 6-1
☐ Alternate Opener, Explorations Transparency 6-1, TE p. 294, and Exploration 6-1
☐ Reaching All Learners TE p. 295
☐ *Know-It Notebook* 6-1

PRACTICE AND APPLY
☐ Example 1: Average: 1, 8–17 Advanced: 3, 8–17
☐ Example 2: Average: 1–5, 8–17 Advanced: 3–4, 6–17

REACHING ALL LEARNERS – Differentiated Instruction for students with

Developing Knowledge	On-level Knowledge	Advanced Knowledge	English Language Development
☐ Multiple Representations TE p. 295	☐ Multiple Representations TE p. 295	☐ Multiple Representations TE p. 295	☐ Multiple Representations TE p. 295
☐ Practice A 6-1 CRB	☐ Practice B 6-1 CRB	☐ Practice C 6-1 CRB	☐ Practice A, B, or C 6-1 CRB
☐ Reteach 6-1 CRB	☐ Puzzles, Twisters & Teasers 6-1 CRB	☐ Challenge 6-1 CRB	☐ *Success for ELL* 6-1
☐ Homework Help Online Keyword: MR7 6-1	☐ Homework Help Online Keyword: MR7 6-1	☐ Homework Help Online Keyword: MR7 6-1	☐ Homework Help Online Keyword: MR7 6-1
☐ *Lesson Tutorial Video* 6-1	☐ *Lesson Tutorial Video* 6-1	☐ *Lesson Tutorial Video* 6-1	☐ *Lesson Tutorial Video* 6-1
☐ *Reading Strategies* 6-1 CRB	☐ *Problem Solving* 6-1 CRB	☐ *Problem Solving* 6-1 CRB	☐ *Reading Strategies* 6-1 CRB
☐ *Questioning Strategies* pp. 86–87			
☐ *IDEA Works!* 6-1			☐ *Multilingual Glossary*

ASSESSMENT
☐ Lesson Quiz, TE p. 296 and DT 6-1   ☐ State-Specific Test Prep Online Keyword: MR7 TestPrep

Teacher's Name _____ Class _____ Date _____

Lesson Plan 6-2
Mean, Median, Mode, and Range pp. 298–301   Day _____

Objective Students find the range, mean, median, and mode of a data set.

> **NCTM Standards:** Compute fluently and make reasonable estimates; Select and use appropriate statistical methods to analyze data.

Pacing
- [ ] 45-minute Classes: 1 day   - [ ] 90-minute Classes: 1/2 day   - [ ] Other _____

WARM UP
- [ ] Warm Up TE p. 298 and Daily Transparency 6-2
- [ ] Problem of the Day TE p. 298 and Daily Transparency 6-2
- [ ] Countdown to Testing Transparency Week 12

TEACH
- [ ] *Lesson Presentation CD-ROM* 6-2
- [ ] Alternate Opener, Explorations Transparency 6-2, TE p. 298, and Exploration 6-2
- [ ] Reaching All Learners TE p. 299
- [ ] *Hands-On Lab Activities* 6-2
- [ ] *Know-It Notebook* 6-2

PRACTICE AND APPLY
- [ ] Example 1: Average: 1, 19–26 Advanced: 3, 19–26
- [ ] Example 2: Average: 1–14, 19–26 Advanced: 3–5, 9–14, 16–26

REACHING ALL LEARNERS – Differentiated Instruction for students with

Developing Knowledge	On-level Knowledge	Advanced Knowledge	English Language Development
☐ Diversity TE p. 299	☐ Diversity TE p. 299	☐ Diversity TE p. 299	☐ Diversity TE p. 299
☐ Practice A 6-2 CRB	☐ Practice B 6-2 CRB	☐ Practice C 6-2 CRB	☐ Practice A, B, or C 6-2 CRB
☐ Reteach 6-2 CRB	☐ Puzzles, Twisters & Teasers 6-2 CRB	☐ Challenge 6-2 CRB	☐ *Success for ELL* 6-2
☐ Homework Help Online Keyword: MR7 6-2	☐ Homework Help Online Keyword: MR7 6-2	☐ Homework Help Online Keyword: MR7 6-2	☐ Homework Help Online Keyword: MR7 6-2
☐ *Lesson Tutorial Video* 6-2	☐ *Lesson Tutorial Video* 6-2	☐ *Lesson Tutorial Video* 6-2	☐ *Lesson Tutorial Video* 6-2
☐ Reading Strategies 6-2 CRB	☐ Problem Solving 6-2 CRB	☐ Problem Solving 6-2 CRB	☐ Reading Strategies 6-2 CRB
☐ Questioning Strategies pp. 88–89			☐ Lesson Vocabulary SE p. 298
☐ *IDEA Works!* 6-2			☐ *Multilingual Glossary*

ASSESSMENT
- [ ] Lesson Quiz, TE p. 301 and DT 6-2   - [ ] State-Specific Test Prep Online Keyword: MR7 TestPrep

Holt Mathematics

Teacher's Name _____ Class _____ Date _____

Lesson Plan 6-3
Additional Data and Outliers pp. 302–305 Day _____

Objective Students learn the effect of additional data and outliers.

> **NCTM Standards:** Select and use appropriate statistical methods to analyze data.

Pacing
☐ 45-minute Classes: 1 day   ☐ 90-minute Classes: 1/2 day   ☐ Other_____

WARM UP
☐ Warm Up TE p. 302 and Daily Transparency 6-3
☐ Problem of the Day TE p. 302 and Daily Transparency 6-3
☐ Countdown to Testing Transparency Week 12

TEACH
☐ Lesson Presentation CD-ROM 6-3
☐ Alternate Opener, Explorations Transparency 6-3, TE p. 302, and Exploration 6-3
☐ Reaching All Learners TE p. 303
☐ *Know-It Notebook* 6-3

PRACTICE AND APPLY
☐ Example 1: Average: 1, 12–17 Advanced: 4, 12–17
☐ Example 2: Average: 1–2, 7–8, 12–17 Advanced: 4–5, 7, 9, 12–17
☐ Example 3: Average: 1–9, 12–17 Advanced: 4–17

REACHING ALL LEARNERS – Differentiated Instruction for students with

Developing Knowledge	On-level Knowledge	Advanced Knowledge	English Language Development
☐ Cooperative Learning TE p. 303	☐ Cooperative Learning TE p. 303	☐ Cooperative Learning TE p. 303	☐ Cooperative Learning TE p. 303
☐ Practice A 6-3 CRB	☐ Practice B 6-3 CRB	☐ Practice C 6-3 CRB	☐ Practice A, B, or C 6-3 CRB
☐ Reteach 6-3 CRB	☐ Puzzles, Twisters & Teasers 6-3 CRB	☐ Challenge 6-3 CRB	☐ *Success for ELL* 6-3
☐ Homework Help Online Keyword: MR7 6-3	☐ Homework Help Online Keyword: MR7 6-3	☐ Homework Help Online Keyword: MR7 6-3	☐ Homework Help Online Keyword: MR7 6-3
☐ *Lesson Tutorial Video* 6-3	☐ *Lesson Tutorial Video* 6-3	☐ *Lesson Tutorial Video* 6-3	☐ *Lesson Tutorial Video* 6-3
☐ Reading Strategies 6-3 CRB	☐ Problem Solving 6-3 CRB	☐ Problem Solving 6-3 CRB	☐ Reading Strategies 6-3 CRB
☐ *Questioning Strategies* pp. 90–91			☐ Lesson Vocabulary SE p. 302
☐ *IDEA Works!* 6-3			☐ *Multilingual Glossary*

ASSESSMENT
☐ Lesson Quiz, TE p. 305 and DT 6-3   ☐ State-Specific Test Prep Online Keyword: MR7 TestPrep

Teacher's Name _____ Class _____ Date _____

Lesson Plan 6-4
Bar Graphs pp. 308–311   Day _____

Objective Students display and analyze data in bar graphs.

> **NCTM Standards:** Formulate questions that can be addressed with data and collect, organize, and display relevant data to answer them; Select and use appropriate statistical methods to analyze data.

Pacing
☐ 45-minute Classes: 1 day   ☐ 90-minute Classes: 1/2 day   ☐ Other_____

WARM UP
☐ Warm Up TE p. 308 and Daily Transparency 6-4
☐ Problem of the Day TE p. 308 and Daily Transparency 6-4
☐ Countdown to Testing Transparency Week 12

TEACH
☐ Lesson Presentation CD-ROM 6-4
☐ Alternate Opener, Explorations Transparency 6-4, TE p. 308, and Exploration 6-4
☐ Reaching All Learners TE p. 309
☐ Teaching Transparency 6-4
☐ *Hands-On Lab Activities* 6-4
☐ *Know-It Notebook* 6-4

PRACTICE AND APPLY
☐ Example 1: Average: 1–2, 16–21   Advanced: 5–6, 16–21
☐ Example 2: Average: 1–3, 16–21   Advanced: 5–7, 16–21
☐ Example 3: Average: 1–10, 16–21   Advanced: 5–21

REACHING ALL LEARNERS – Differentiated Instruction for students with

Developing Knowledge	On-level Knowledge	Advanced Knowledge	English Language Development
☐ Concrete Manipulatives TE p. 309	☐ Concrete Manipulatives TE p. 309	☐ Concrete Manipulatives TE p. 309	☐ Concrete Manipulatives TE p. 309
☐ Practice A 6-4 CRB	☐ Practice B 6-4 CRB	☐ Practice C 6-4 CRB	☐ Practice A, B, or C 6-4 CRB
☐ Reteach 6-4 CRB	☐ Puzzles, Twisters & Teasers 6-4 CRB	☐ Challenge 6-4 CRB	☐ *Success for ELL* 6-4
☐ Homework Help Online Keyword: MR7 6-4	☐ Homework Help Online Keyword: MR7 6-4	☐ Homework Help Online Keyword: MR7 6-4	☐ Homework Help Online Keyword: MR7 6-4
☐ *Lesson Tutorial Video* 6-4	☐ *Lesson Tutorial Video* 6-4	☐ *Lesson Tutorial Video* 6-4	☐ *Lesson Tutorial Video* 6-4
☐ Reading Strategies 6-4 CRB	☐ Problem Solving 6-4 CRB	☐ Problem Solving 6-4 CRB	☐ Reading Strategies 6-4 CRB
☐ *Questioning Strategies* pp. 92–93	☐ Critical Thinking TE p. 309	☐ Critical Thinking TE p. 309	☐ Lesson Vocabulary SE p. 308
☐ *IDEA Works!* 6-4			☐ *Multilingual Glossary*

ASSESSMENT
☐ Lesson Quiz, TE p. 311 and DT 6-4   ☐ State-Specific Test Prep Online Keyword: MR7 TestPrep

Copyright © Holt, Rinehart and Winston.
All rights reserved.

Holt Mathematics

Teacher's Name _____ Class _____ Date _____

Lesson Plan 6-5
Line Plots, Frequency Tables, and Histograms pp. 314–317   Day _____

Objective Students record and organize data in line plots, frequency tables, and histograms.

> **NCTM Standards:** Select and use appropriate statistical methods to analyze data; Create and use representations to organize, record, and communicate mathematical ideas.

Pacing
- ☐ 45-minute Classes: 1 day   ☐ 90-minute Classes: 1/2 day   ☐ Other_____

WARM UP
- ☐ Warm Up TE p. 314 and Daily Transparency 6-5
- ☐ Problem of the Day TE p. 314 and Daily Transparency 6-5
- ☐ Countdown to Testing Transparency Week 12

TEACH
- ☐ Lesson Presentation CD-ROM 6-5
- ☐ Alternate Opener, Explorations Transparency 6-5, TE p. 314, and Exploration 6-5
- ☐ Reaching All Learners TE p. 315
- ☐ Hands-On Lab Activities 6-5
- ☐ Know-It Notebook 6-5

PRACTICE AND APPLY
- ☐ Example 1: Average: 1, 17–25 Advanced: 5, 17–25
- ☐ Example 2: Average: 1–2, 10, 17–25 Advanced: 5–6, 10, 17–25
- ☐ Example 3: Average: 1–3, 10–11, 17–25 Advanced: 5–7, 10–11, 17–25
- ☐ Example 4: Average: 1–12, 17–25 Advanced: 5–11, 14–25

REACHING ALL LEARNERS – Differentiated Instruction for students with

Developing Knowledge	On-level Knowledge	Advanced Knowledge	English Language Development
☐ Cooperative Learning TE p. 315	☐ Cooperative Learning TE p. 315	☐ Cooperative Learning TE p. 315	☐ Cooperative Learning TE p. 315
☐ Practice A 6-5 CRB	☐ Practice B 6-5 CRB	☐ Practice C 6-5 CRB	☐ Practice A, B, or C 6-5 CRB
☐ Reteach 6-5 CRB	☐ Puzzles, Twisters & Teasers 6-5 CRB	☐ Challenge 6-5 CRB	☐ *Success for ELL* 6-5
☐ Homework Help Online Keyword: MR7 6-5	☐ Homework Help Online Keyword: MR7 6-5	☐ Homework Help Online Keyword: MR7 6-5	☐ Homework Help Online Keyword: MR7 6-5
☐ *Lesson Tutorial Video* 6-5	☐ *Lesson Tutorial Video* 6-5	☐ *Lesson Tutorial Video* 6-5	☐ *Lesson Tutorial Video* 6-5
☐ Reading Strategies 6-5 CRB	☐ Problem Solving 6-5 CRB	☐ Problem Solving 6-5 CRB	☐ Reading Strategies 6-5 CRB
☐ *Questioning Strategies* pp. 94–95			☐ Lesson Vocabulary SE p. 314
☐ *IDEA Works!* 6-5			☐ *Multilingual Glossary*

ASSESSMENT
- ☐ Lesson Quiz, TE p. 317 and DT 6-5   ☐ State-Specific Test Prep Online Keyword: MR7 TestPrep

Teacher's Name _____ Class _____ Date _____

Lesson Plan 6-6
Ordered Pairs pp. 319–321   Day _____

Objective Students graph ordered pairs on a coordinate grid.

> **NCTM Standards:** Understand patterns, relations, and functions; Specify locations and describe spatial relationships using coordinate geometry and other representational systems.

Pacing
- ☐ 45-minute Classes: 1 day   ☐ 90-minute Classes: 1/2 day   ☐ Other_____

WARM UP
- ☐ Warm Up TE p. 319 and Daily Transparency 6-6
- ☐ Problem of the Day TE p. 319 and Daily Transparency 6-6
- ☐ Countdown to Testing Transparency Week 13

TEACH
- ☐ Lesson Presentation CD-ROM 6-6
- ☐ Alternate Opener, Explorations Transparency 6-6, TE p. 319, and Exploration 6-6
- ☐ Reaching All Learners TE p. 320
- ☐ Teaching Transparency 6-6
- ☐ *Hands-On Lab Activities* 6-6
- ☐ *Know-It Notebook* 6-6

PRACTICE AND APPLY
- ☐ Example 1: Average: 1–6, 29–34, 39–47   Advanced: 11–16, 29–34, 39–47
- ☐ Example 2: Average: 1–26, 39–47   Advanced: 11–32, 37–47

REACHING ALL LEARNERS – Differentiated Instruction for students with

Developing Knowledge	On-level Knowledge	Advanced Knowledge	English Language Development
☐ Kinesthetic Experience TE p. 320	☐ Kinesthetic Experience TE p. 320	☐ Kinesthetic Experience TE p. 320	☐ Kinesthetic Experience TE p. 320
☐ Practice A 6-6 CRB	☐ Practice B 6-6 CRB	☐ Practice C 6-6 CRB	☐ Practice A, B, or C 6-6 CRB
☐ Reteach 6-6 CRB	☐ Puzzles, Twisters & Teasers 6-6 CRB	☐ Challenge 6-6 CRB	☐ *Success for ELL* 6-6
☐ Homework Help Online Keyword: MR7 6-6	☐ Homework Help Online Keyword: MR7 6-6	☐ Homework Help Online Keyword: MR7 6-6	☐ Homework Help Online Keyword: MR7 6-6
☐ *Lesson Tutorial Video* 6-6	☐ *Lesson Tutorial Video* 6-6	☐ *Lesson Tutorial Video* 6-6	☐ *Lesson Tutorial Video* 6-6
☐ Reading Strategies 6-6 CRB	☐ Problem Solving 6-6 CRB	☐ Problem Solving 6-6 CRB	☐ Reading Strategies 6-6 CRB
☐ *Questioning Strategies* pp. 96–97			☐ Lesson Vocabulary SE p. 319
☐ *IDEA Works!* 6-6			☐ *Multilingual Glossary*

ASSESSMENT
- ☐ Lesson Quiz, TE p. 321 and DT 6-6   ☐ State-Specific Test Prep Online Keyword: MR7 TestPrep

Teacher's Name _____ Class _____ Date _____

Lesson Plan 6-7
Line Graphs pp. 322–325   Day _____

Objective Students display and analyze data in line graphs.

> **NCTM Standards:** Select and use appropriate statistical methods to analyze data.

Pacing
☐ 45-minute Classes: 1 day   ☐ 90-minute Classes: 1/2 day   ☐ Other_____

WARM UP
☐ Warm Up TE p. 322 and Daily Transparency 6-7
☐ Problem of the Day TE p. 322 and Daily Transparency 6-7
☐ Countdown to Testing Transparency Week 13

TEACH
☐ Lesson Presentation CD-ROM 6-7
☐ Alternate Opener, Explorations Transparency 6-7, TE p. 322, and Exploration 6-7
☐ Reaching All Learners TE p. 323
☐ Teaching Transparency 6-7
☐ *Technology Lab Activities* 6-7
☐ *Know-It Notebook* 6-7

PRACTICE AND APPLY
☐ Example 1: Average: 1, 15–21 Advanced: 6, 15–21
☐ Example 2: Average: 1–4, 15–21 Advanced: 6–8, 15–21
☐ Example 3: Average: 1–8, 12–21 Advanced: 6–21

REACHING ALL LEARNERS – Differentiated Instruction for students with

Developing Knowledge	On-level Knowledge	Advanced Knowledge	English Language Development
☐ Inclusion TE p. 323	☐ Cognitive Strategies TE p. 323	☐ Cognitive Strategies TE p. 323	☐ Cognitive Strategies TE p. 323
☐ Practice A 6-7 CRB	☐ Practice B 6-7 CRB	☐ Practice C 6-7 CRB	☐ Practice A, B, or C 6-7 CRB
☐ Reteach 6-7 CRB	☐ Puzzles, Twisters & Teasers 6-7 CRB	☐ Challenge 6-7 CRB	☐ *Success for ELL* 6-7
☐ Homework Help Online Keyword: MR7 6-7	☐ Homework Help Online Keyword: MR7 6-7	☐ Homework Help Online Keyword: MR7 6-7	☐ Homework Help Online Keyword: MR7 6-7
☐ *Lesson Tutorial Video* 6-7	☐ *Lesson Tutorial Video* 6-7	☐ *Lesson Tutorial Video* 6-7	☐ *Lesson Tutorial Video* 6-7
☐ Reading Strategies 6-7 CRB	☐ Problem Solving 6-7 CRB	☐ Problem Solving 6-7 CRB	☐ Reading Strategies 6-7 CRB
☐ *Questioning Strategies* pp. 98–99			☐ Lesson Vocabulary SE p. 322
☐ *IDEA Works!* 6-7			☐ *Multilingual Glossary*

ASSESSMENT
☐ Lesson Quiz, TE p. 325 and DT 6-7   ☐ State-Specific Test Prep Online Keyword: MR7 TestPrep

Teacher's Name _____ Class _____ Date _____

Lesson Plan 6-8
Misleading Graphs pp. 326–329   Day _____

Objective Students recognize misleading graphs.

> **NCTM Standards:** Develop and evaluate inferences and predictions that are based on data.

Pacing
☐ 45-minute Classes: 1 day   ☐ 90-minute Classes: 1/2 day   ☐ Other _____

WARM UP
☐ Warm Up TE p. 326 and Daily Transparency 6-8
☐ Problem of the Day TE p. 326 and Daily Transparency 6-8
☐ Countdown to Testing Transparency Week 13

TEACH
☐ Lesson Presentation CD-ROM 6-8
☐ Alternate Opener, Explorations Transparency 6-8, TE p. 326, and Exploration 6-8
☐ Reaching All Learners TE p. 327
☐ Teaching Transparency 6-8
☐ *Know-It Notebook* 6-8

PRACTICE AND APPLY
☐ Example 1: Average: 1–2, 16–24   Advanced: 5–6, 16–24
☐ Example 2: Average: 1–9, 16–24   Advanced: 5–9, 13–24

REACHING ALL LEARNERS – Differentiated Instruction for students with

Developing Knowledge	On-level Knowledge	Advanced Knowledge	English Language Development
☐ Inclusion TE p. 327	☐ Cooperative Learning TE p. 327	☐ Cooperative Learning TE p. 327	☐ Cooperative Learning TE p. 327
☐ Practice A 6-8 CRB	☐ Practice B 6-8 CRB	☐ Practice C 6-8 CRB	☐ Practice A, B, or C 6-8 CRB
☐ Reteach 6-8 CRB	☐ Puzzles, Twisters & Teasers 6-8 CRB	☐ Challenge 6-8 CRB	☐ *Success for ELL* 6-8
☐ Homework Help Online Keyword: MR7 6-8	☐ Homework Help Online Keyword: MR7 6-8	☐ Homework Help Online Keyword: MR7 6-8	☐ Homework Help Online Keyword: MR7 6-8
☐ *Lesson Tutorial Video* 6-8	☐ *Lesson Tutorial Video* 6-8	☐ *Lesson Tutorial Video* 6-8	☐ *Lesson Tutorial Video* 6-8
☐ Reading Strategies 6-8 CRB	☐ Problem Solving 6-8 CRB	☐ Problem Solving 6-8 CRB	☐ Reading Strategies 6-8 CRB
☐ *Questioning Strategies* pp. 100–101			
☐ *IDEA Works!* 6-8			☐ *Multilingual Glossary*

ASSESSMENT
☐ Lesson Quiz, TE p. 329 and DT 6-8   ☐ State-Specific Test Prep Online   Keyword: MR7 TestPrep

Holt Mathematics

Teacher's Name _____ Class _____ Date _____

Lesson Plan 6-9
Stem-and-Leaf Plots pp. 330–332   Day _____

Objective Students make and analyze stem-and-leaf plots.

> **NCTM Standards:** Select and use appropriate statistical methods to analyze data.

Pacing
☐ 45-minute Classes: 1 day   ☐ 90-minute Classes: 1/2 day   ☐ Other _____

WARM UP
☐ Warm Up TE p. 330 and Daily Transparency 6-9
☐ Problem of the Day TE p. 330 and Daily Transparency 6-9
☐ Countdown to Testing Transparency Week 13

TEACH
☐ Lesson Presentation CD-ROM 6-9
☐ Alternate Opener, Explorations Transparency 6-9, TE p. 330, and Exploration 6-9
☐ Reaching All Learners TE p. 331
☐ *Know-It Notebook* 6-9

PRACTICE AND APPLY
☐ Example 1: Average: 1, 17, 20–28 Advanced: 8, 17, 20–28
☐ Example 2: Average: 1–8, 15–16, 20–28 Advanced: 8–14, 17–28

REACHING ALL LEARNERS – Differentiated Instruction for students with

Developing Knowledge	On-level Knowledge	Advanced Knowledge	English Language Development
☐ Critical Thinking TE p. 331	☐ Critical Thinking TE p. 331	☐ Critical Thinking TE p. 331	☐ Critical Thinking TE p. 331
☐ Practice A 6-9 CRB	☐ Practice B 6-9 CRB	☐ Practice C 6-9 CRB	☐ Practice A, B, or C 6-9 CRB
☐ Reteach 6-9 CRB	☐ Puzzles, Twisters & Teasers 6-9 CRB	☐ Challenge 6-9 CRB	☐ *Success for ELL* 6-9
☐ Homework Help Online Keyword: MR7 6-9	☐ Homework Help Online Keyword: MR7 6-9	☐ Homework Help Online Keyword: MR7 6-9	☐ Homework Help Online Keyword: MR7 6-9
☐ *Lesson Tutorial Video* 6-9	☐ *Lesson Tutorial Video* 6-9	☐ *Lesson Tutorial Video* 6-9	☐ *Lesson Tutorial Video* 6-9
☐ Reading Strategies 6-9 CRB	☐ Problem Solving 6-9 CRB	☐ Problem Solving 6-9 CRB	☐ Reading Strategies 6-9 CRB
☐ *Questioning Strategies* pp. 102–103			☐ Lesson Vocabulary SE p. 330
☐ *IDEA Works!* 6-9			☐ *Multilingual Glossary*

ASSESSMENT
☐ Lesson Quiz, TE p. 332 and DT 6-9   ☐ State-Specific Test Prep Online Keyword: MR7 TestPrep

Teacher's Name _____ Class _____ Date _____

Lesson Plan 6-10
Choosing an Appropriate Display pp. 333–335   Day _____

Objective Students choose an appropriate way to display data.

NCTM Standards: Select and use appropriate statistical methods to analyze data.

Pacing
☐ 45-minute Classes: 1 day   ☐ 90-minute Classes: 1/2 day   ☐ Other_____

WARM UP
☐ Warm Up TE p. 333 and Daily Transparency 6-10
☐ Problem of the Day TE p. 333 and Daily Transparency 6-10
☐ Countdown to Testing Transparency Week 13

TEACH
☐ Lesson Presentation CD-ROM 6-10
☐ Alternate Opener, Explorations Transparency 6-10, TE p. 333, and Exploration 6-10
☐ Reaching All Learners TE p. 334
☐ *Hands-On Lab Activities* 6-10
☐ *Know-It Notebook* 6-10

PRACTICE AND APPLY
☐ Example 1: Average: 1, 3–4, 8–14 Advanced: 2, 5–14

REACHING ALL LEARNERS – Differentiated Instruction for students with

Developing Knowledge	On-level Knowledge	Advanced Knowledge	English Language Development
☐ Cooperative Learning TE p. 334	☐ Cooperative Learning TE p. 334	☐ Cooperative Learning TE p. 334	☐ Cooperative Learning TE p. 334
☐ Practice A 6-10 CRB	☐ Practice B 6-10 CRB	☐ Practice C 6-10 CRB	☐ Practice A, B, or C 6-10 CRB
☐ Reteach 6-10 CRB	☐ Puzzles, Twisters & Teasers 6-10 CRB	☐ Challenge 6-10 CRB	☐ *Success for ELL* 6-10
☐ Homework Help Online Keyword: MR7 6-10	☐ Homework Help Online Keyword: MR7 6-10	☐ Homework Help Online Keyword: MR7 6-10	☐ Homework Help Online Keyword: MR7 6-10
☐ *Lesson Tutorial Video* 6-10	☐ *Lesson Tutorial Video* 6-10	☐ *Lesson Tutorial Video* 6-10	☐ *Lesson Tutorial Video* 6-10
☐ Reading Strategies 6-10 CRB	☐ Problem Solving 6-10 CRB	☐ Problem Solving 6-10 CRB	☐ Reading Strategies 6-10 CRB
☐ *Questioning Strategies* pp. 104–105			
☐ *IDEA Works!* 6-10			☐ *Multilingual Glossary*

ASSESSMENT
☐ Lesson Quiz, TE p. 335 and DT 6-10   ☐ State-Specific Test Prep Online Keyword: MR7 TestPrep

Teacher's Name _____ Class _____ Date _____

Lesson Plan 7-1
Ratios and Rates pp. 352–355   Day _____

Objective Students write ratios and rates and find unit rates.

> **NCTM Standards:** Understand numbers, ways of representing numbers, relationships among numbers, and number systems.

Pacing
☐ 45-minute Classes: 1 day ☐ 90-minute Classes: 1/2 day ☐ Other_____

WARM UP
☐ Warm Up TE p. 352 and Daily Transparency 7-1
☐ Problem of the Day TE p. 352 and Daily Transparency 7-1
☐ Countdown to Testing Transparency Week 14

TEACH
☐ Lesson Presentation CD-ROM 7-1
☐ Alternate Opener, Explorations Transparency 7-1, TE p. 352, and Exploration 7-1
☐ Reaching All Learners TE p. 353
☐ Teaching Transparency 7-1
☐ *Know-It Notebook* 7-1

PRACTICE AND APPLY
☐ Example 1: Average: 1–3, 32–41 Advanced: 6–8, 32–41
☐ Example 2: Average: 1–4, 21–24, 32–41 Advanced: 6–9, 21–24, 32–41
☐ Example 3: Average: 1–14, 20–41 Advanced: 6–41

REACHING ALL LEARNERS – Differentiated Instruction for students with

Developing Knowledge	On-level Knowledge	Advanced Knowledge	English Language Development
☐ Cooperative Learning TE p. 353	☐ Cooperative Learning TE p. 353	☐ Cooperative Learning TE p. 353	☐ Cooperative Learning TE p. 353
☐ Practice A 7-1 CRB	☐ Practice B 7-1 CRB	☐ Practice C 7-1 CRB	☐ Practice A, B, or C 7-1 CRB
☐ Reteach 7-1 CRB	☐ Puzzles, Twisters & Teasers 7-1 CRB	☐ Challenge 7-1 CRB	☐ *Success for ELL* 7-1
☐ Homework Help Online Keyword: MR7 7-1	☐ Homework Help Online Keyword: MR7 7-1	☐ Homework Help Online Keyword: MR7 7-1	☐ Homework Help Online Keyword: MR7 7-1
☐ *Lesson Tutorial Video* 7-1	☐ *Lesson Tutorial Video* 7-1	☐ *Lesson Tutorial Video* 7-1	☐ *Lesson Tutorial Video* 7-1
☐ Reading Strategies 7-1 CRB	☐ Problem Solving 7-1 CRB	☐ Problem Solving 7-1 CRB	☐ Reading Strategies 7-1 CRB
☐ *Questioning Strategies* pp. 106–107			☐ *Lesson Vocabulary* SE p. 352
☐ *IDEA Works!* 7-1			☐ *Multilingual Glossary*

ASSESSMENT
☐ Lesson Quiz, TE p. 355 and DT 7-1   ☐ State-Specific Test Prep Online Keyword: MR7 TestPrep

Teacher's Name _____ Class _____ Date _____

Lesson Plan 7-2
Using Tables to Explore Equivalent Ratios and Rates pp. 356–359   Day _____

Objective Students use a table to find equivalent ratios and rates.

> **NCTM Standards:** Understand numbers, ways of representing numbers, relationships among numbers, and number systems.

Pacing
☐ 45-minute Classes: 1 day   ☐ 90-minute Classes: 1/2 day   ☐ Other_____

WARM UP
☐ Warm Up TE p. 356 and Daily Transparency 7-2
☐ Problem of the Day TE p. 356 and Daily Transparency 7-2
☐ Countdown to Testing Transparency Week 14

TEACH
☐ Lesson Presentation CD-ROM 7-2
☐ Alternate Opener, Explorations Transparency 7-2, TE p. 356, and Exploration 7-2
☐ Reaching All Learners TE p. 357
☐ Teaching Transparency 7-2
☐ *Know-It Notebook* 7-2

PRACTICE AND APPLY
☐ Example 1: Average: 1–8, 31–34 Advanced: 10–17, 31–34
☐ Example 2: Average: 1–22, 31–34 Advanced: 10–34

REACHING ALL LEARNERS – Differentiated Instruction for students with

Developing Knowledge	On-level Knowledge	Advanced Knowledge	English Language Development
☐ Modeling TE p. 357	☐ Modeling TE p. 357	☐ Modeling TE p. 357	☐ Modeling TE p. 357
☐ Practice A 7-2 CRB	☐ Practice B 7-2 CRB	☐ Practice C 7-2 CRB	☐ Practice A, B, or C 7-2 CRB
☐ Reteach 7-2 CRB	☐ Puzzles, Twisters & Teasers 7-2 CRB	☐ Challenge 7-2 CRB	☐ *Success for ELL* 7-2
☐ Homework Help Online Keyword: MR7 7-2	☐ Homework Help Online Keyword: MR7 7-2	☐ Homework Help Online Keyword: MR7 7-2	☐ Homework Help Online Keyword: MR7 7-2
☐ Lesson Tutorial Video 7-2	☐ Lesson Tutorial Video 7-2	☐ Lesson Tutorial Video 7-2	☐ Lesson Tutorial Video 7-2
☐ Reading Strategies 7-2 CRB	☐ Problem Solving 7-2 CRB	☐ Problem Solving 7-2 CRB	☐ Reading Strategies 7-2 CRB
☐ Questioning Strategies pp. 108–109			
☐ *IDEA Works!* 7-2			☐ *Multilingual Glossary*

ASSESSMENT
☐ Lesson Quiz, TE p. 359 and DT 7-2   ☐ State-Specific Test Prep Online Keyword: MR7 TestPrep

Teacher's Name _____ Class _____ Date _____

Lesson Plan 7-3
Proportions pp. 362–365   Day _____

Objective Students write and solve proportions.

> **NCTM Standards:** Select and use various types of reasoning and methods of proof; Recognize and use connections among mathematical ideas; Understand how mathematical ideas interconnect and build on one another to produce a coherent whole.

Pacing
☐ 45-minute Classes: 1 day   ☐ 90-minute Classes: 1/2 day   ☐ Other _____

WARM UP
☐ Warm Up TE p. 362 and Daily Transparency 7-3
☐ Problem of the Day TE p. 362 and Daily Transparency 7-3
☐ Countdown to Testing Transparency Week 14

TEACH
☐ Lesson Presentation CD-ROM 7-3
☐ Alternate Opener, Explorations Transparency 7-3, TE p. 362, and Exploration 7-3
☐ Reaching All Learners TE p. 363
☐ Teaching Transparency 7-3
☐ *Know-It Notebook* 7-3

PRACTICE AND APPLY
☐ Example 1: Average: 1, 27–36 Advanced: 7, 27–36
☐ Example 2: Average: 1–5, 13–16, 27–36 Advanced: 7–11, 17–20, 27–36
☐ Example 3: Average: 1–10, 16–36 Advanced: 7–36

REACHING ALL LEARNERS – Differentiated Instruction for students with

Developing Knowledge	On-level Knowledge	Advanced Knowledge	English Language Development
☐ Curriculum Integration TE p. 363	☐ Curriculum Integration TE p. 363	☐ Curriculum Integration TE p. 363	☐ Curriculum Integration TE p. 363
☐ Practice A 7-3 CRB	☐ Practice B 7-3 CRB	☐ Practice C 7-3 CRB	☐ Practice A, B, or C 7-3 CRB
☐ Reteach 7-3 CRB	☐ Puzzles, Twisters & Teasers 7-3 CRB	☐ Challenge 7-3 CRB	☐ *Success for ELL* 7-3
☐ Homework Help Online Keyword: MR7 7-3	☐ Homework Help Online Keyword: MR7 7-3	☐ Homework Help Online Keyword: MR7 7-3	☐ Homework Help Online Keyword: MR7 7-3
☐ *Lesson Tutorial Video* 7-3	☐ *Lesson Tutorial Video* 7-3	☐ *Lesson Tutorial Video* 7-3	☐ *Lesson Tutorial Video* 7-3
☐ Reading Strategies 7-3 CRB	☐ Problem Solving 7-3 CRB	☐ Problem Solving 7-3 CRB	☐ Reading Strategies 7-3 CRB
☐ *Questioning Strategies* pp. 110–111	☐ Concrete Manipulatives TE p. 363	☐ Concrete Manipulatives TE p. 363	☐ Lesson Vocabulary SE p. 362
☐ *IDEA Works!* 7-3			☐ *Multilingual Glossary*

ASSESSMENT
☐ Lesson Quiz, TE p. 365 and DT 7-3   ☐ State-Specific Test Prep Online Keyword: MR7 TestPrep

Teacher's Name _____ Class _____ Date _____

Lesson Plan 7-4
Similar Figures pp. 366–369   Day _____

Objective Students use ratios to identify similar figures.

> **NCTM Standards:** Compute fluently and make reasonable estimates; Use visualization, spatial reasoning, and geometric modeling to solve problems; Recognize and use connections among mathematical ideas.

Pacing
- [ ] 45-minute Classes: 1 day   [ ] 90-minute Classes: 1/2 day   [ ] Other_____

WARM UP
- [ ] Warm Up TE p. 366 and Daily Transparency 7-4
- [ ] Problem of the Day TE p. 366 and Daily Transparency 7-4
- [ ] Countdown to Testing Transparency Week 14

TEACH
- [ ] Lesson Presentation CD-ROM 7-4
- [ ] Alternate Opener, Explorations Transparency 7-4, TE p. 366, and Exploration 7-4
- [ ] Reaching All Learners TE p. 367
- [ ] Teaching Transparency 7-4
- [ ] *Hands-On Lab Activities* 7-4
- [ ] *Know-It Notebook* 7-4

PRACTICE AND APPLY
- [ ] Example 1: Average: 1, 8, 16–23 Advanced: 3, 9, 16–23
- [ ] Example 2: Average: 1–4, 7–23 Advanced: 3–23

REACHING ALL LEARNERS – Differentiated Instruction for students with

Developing Knowledge	On-level Knowledge	Advanced Knowledge	English Language Development
[ ] Inclusion TE p. 367	[ ] Inclusion TE p. 367	[ ] Inclusion TE p. 367	[ ] Inclusion TE p. 367
[ ] Practice A 7-4 CRB	[ ] Practice B 7-4 CRB	[ ] Practice C 7-4 CRB	[ ] Practice A, B, or C 7-4 CRB
[ ] Reteach 7-4 CRB	[ ] Puzzles, Twisters & Teasers 7-4 CRB	[ ] Challenge 7-4 CRB	[ ] *Success for ELL* 7-4
[ ] Homework Help Online Keyword: MR7 7-4	[ ] Homework Help Online Keyword: MR7 7-4	[ ] Homework Help Online Keyword: MR7 7-4	[ ] Homework Help Online Keyword: MR7 7-4
[ ] *Lesson Tutorial Video* 7-4	[ ] *Lesson Tutorial Video* 7-4	[ ] *Lesson Tutorial Video* 7-4	[ ] *Lesson Tutorial Video* 7-4
[ ] Reading Strategies 7-4 CRB	[ ] Problem Solving 7-4 CRB	[ ] Problem Solving 7-4 CRB	[ ] Reading Strategies 7-4 CRB
[ ] *Questioning Strategies* pp. 112–113			[ ] Lesson Vocabulary SE p. 366
[ ] *IDEA Works!* 7-4			[ ] *Multilingual Glossary*

ASSESSMENT
- [ ] Lesson Quiz, TE p. 369 and DT 7-4   [ ] State-Specific Test Prep Online Keyword: MR7 TestPrep

Teacher's Name _____ Class _____ Date _____

Lesson Plan 7-5
Indirect Measurement pp. 370–372 Day _____

Objective Students use proportions and similar figures to find unknown measures.

> **NCTM Standards:** Compute fluently and make reasonable estimates; Use visualization, spatial reasoning, and geometric modeling to solve problems; Apply appropriate techniques, tools, and formulas to determine measurements.

Pacing
☐ 45-minute Classes: 1 day ☐ 90-minute Classes: 1/2 day ☐ Other_____

WARM UP
☐ Warm Up TE p. 370 and Daily Transparency 7-5
☐ Problem of the Day TE p. 370 and Daily Transparency 7-5
☐ Countdown to Testing Transparency Week 15

TEACH
☐ Lesson Presentation CD-ROM 7-5
☐ Alternate Opener, Explorations Transparency 7-5, TE p. 370, and Exploration 7-5
☐ Reaching All Learners TE p. 371
☐ Teaching Transparency 7-5
☐ *Technology Lab Activities* 7-5
☐ *Know-It Notebook* 7-5

PRACTICE AND APPLY
☐ Example 1: Average: 1, 5, 10–18 Advanced: 3, 6, 10–18
☐ Example 2: Average: 1–5, 8–18 Advanced: 3–18

REACHING ALL LEARNERS – Differentiated Instruction for students with

Developing Knowledge	On-level Knowledge	Advanced Knowledge	English Language Development
☐ Diversity TE p. 371	☐ Diversity TE p. 371	☐ Diversity TE p. 371	☐ Diversity TE p. 371
☐ Practice A 7-5 CRB	☐ Practice B 7-5 CRB	☐ Practice C 7-5 CRB	☐ Practice A, B, or C 7-5 CRB
☐ Reteach 7-5 CRB	☐ Puzzles, Twisters & Teasers 7-5 CRB	☐ Challenge 7-5 CRB	☐ *Success for ELL* 7-5
☐ Homework Help Online Keyword: MR7 7-5	☐ Homework Help Online Keyword: MR7 7-5	☐ Homework Help Online Keyword: MR7 7-5	☐ Homework Help Online Keyword: MR7 7-5
☐ *Lesson Tutorial Video* 7-5	☐ *Lesson Tutorial Video* 7-5	☐ *Lesson Tutorial Video* 7-5	☐ *Lesson Tutorial Video* 7-5
☐ Reading Strategies 7-5 CRB	☐ Problem Solving 7-5 CRB	☐ Problem Solving 7-5 CRB	☐ Reading Strategies 7-5 CRB
☐ *Questioning Strategies* pp. 114–115			☐ Lesson Vocabulary SE p. 370
☐ *IDEA Works!* 7-5			☐ *Multilingual Glossary*

ASSESSMENT
☐ Lesson Quiz, TE p. 372 and DT 7-5   ☐ State-Specific Test Prep Online Keyword: MR7 TestPrep

Teacher's Name _____ Class _____ Date _____

Lesson Plan 7-6
Scale Drawings and Maps pp. 374–377   Day _____

Objective Students read and use map scales and scale drawings.

> **NCTM Standards:** Apply appropriate techniques, tools, and formulas to determine measurements; Recognize and apply mathematics in contexts outside of mathematics; Use representations to model and interpret physical, social, and mathematical phenomena.

Pacing
- ☐ 45-minute Classes: 1 day   ☐ 90-minute Classes: 1/2 day   ☐ Other _____

WARM UP
- ☐ Warm Up TE p. 374 and Daily Transparency 7-6
- ☐ Problem of the Day TE p. 374 and Daily Transparency 7-6
- ☐ Countdown to Testing Transparency Week 15

TEACH
- ☐ Lesson Presentation CD-ROM 7-6
- ☐ Alternate Opener, Explorations Transparency 7-6, TE p. 374, and Exploration 7-6
- ☐ Reaching All Learners TE p. 375
- ☐ Teaching Transparency 7-6
- ☐ *Hands-On Lab Activities* 7-6
- ☐ *Know-It Notebook* 7-6

PRACTICE AND APPLY
- ☐ Example 1: Average: 1, 8, 16–27 Advanced: 4, 8, 16–27
- ☐ Example 2: Average: 1–6, 8–27 Advanced: 4–27

REACHING ALL LEARNERS – Differentiated Instruction for students with

Developing Knowledge	On-level Knowledge	Advanced Knowledge	English Language Development
☐ Curriculum Integration TE p. 375	☐ Curriculum Integration TE p. 375	☐ Curriculum Integration TE p. 375	☐ Curriculum Integration TE p. 375
☐ Practice A 7-6 CRB	☐ Practice B 7-6 CRB	☐ Practice C 7-6 CRB	☐ Practice A, B, or C 7-6 CRB
☐ Reteach 7-6 CRB	☐ Puzzles, Twisters & Teasers 7-6 CRB	☐ Challenge 7-6 CRB	☐ *Success for ELL* 7-6
☐ Homework Help Online Keyword: MR7 7-6	☐ Homework Help Online Keyword: MR7 7-6	☐ Homework Help Online Keyword: MR7 7-6	☐ Homework Help Online Keyword: MR7 7-6
☐ *Lesson Tutorial Video* 7-6	☐ *Lesson Tutorial Video* 7-6	☐ *Lesson Tutorial Video* 7-6	☐ *Lesson Tutorial Video* 7-6
☐ Reading Strategies 7-6 CRB	☐ Problem Solving 7-6 CRB	☐ Problem Solving 7-6 CRB	☐ Reading Strategies 7-6 CRB
☐ *Questioning Strategies* pp. 116–117	☐ Critical Thinking TE p. 375	☐ Critical Thinking TE p. 375	☐ Lesson Vocabulary SE p. 374
☐ *IDEA Works!* 7-6			☐ *Multilingual Glossary*

ASSESSMENT
- ☐ Lesson Quiz, TE p. 377 and DT 7-6   ☐ State-Specific Test Prep Online Keyword: MR7 TestPrep

Holt Mathematics

Teacher's Name _____ Class _____ Date _____

Lesson Plan 7-7
Percents pp. 381–384   Day _____

Objective Students write percents as decimals and as fractions.

> **NCTM Standards:** Understand numbers, ways of representing numbers, relationships among numbers, and number systems; Compute fluently and make reasonable estimates; Select, apply, and translate among mathematical representations to solve problems.

Pacing
☐ 45-minute Classes: 1 day   ☐ 90-minute Classes: 1/2 day   ☐ Other_____

WARM UP
☐ Warm Up TE p. 381 and Daily Transparency 7-7
☐ Problem of the Day TE p. 381 and Daily Transparency 7-7
☐ Countdown to Testing Transparency Week 15

TEACH
☐ Lesson Presentation CD-ROM 7-7
☐ Alternate Opener, Explorations Transparency 7-7, TE p. 381, and Exploration 7-7
☐ Reaching All Learners TE p. 382
☐ Teaching Transparency 7-7
☐ *Hands-On Lab Activities* 7-7
☐ *Know-It Notebook* 7-7

PRACTICE AND APPLY
☐ Example 1: Average: 1–3, 51, 58–64 Advanced: 12–15, 51, 58–64
☐ Example 2: Average: 1–6, 50–51, 58–64 Advanced: 12–23, 58–64
☐ Example 3: Average: 1–7, 49–51, 58–64 Advanced: 12–24, 58–64
☐ Example 4: Average: 1–10, 34–51, 58–64 Advanced: 12–32, 58–64
☐ Example 5: Average: 1–11, 22–64 Advanced: 12–64

REACHING ALL LEARNERS – Differentiated Instruction for students with

Developing Knowledge	On-level Knowledge	Advanced Knowledge	English Language Development
☐ Kinesthetic Experience TE p. 382	☐ Kinesthetic Experience TE p. 382	☐ Kinesthetic Experience TE p. 382	☐ Kinesthetic Experience TE p. 382
☐ Practice A 7-7 CRB	☐ Practice B 7-7 CRB	☐ Practice C 7-7 CRB	☐ Practice A, B, or C 7-7 CRB
☐ Reteach 7-7 CRB	☐ Puzzles, Twisters & Teasers 7-7 CRB	☐ Challenge 7-7 CRB	☐ *Success for ELL* 7-7
☐ Homework Help Online Keyword: MR7 7-7	☐ Homework Help Online Keyword: MR7 7-7	☐ Homework Help Online Keyword: MR7 7-7	☐ Homework Help Online Keyword: MR7 7-7
☐ *Lesson Tutorial Video* 7-7	☐ *Lesson Tutorial Video* 7-7	☐ *Lesson Tutorial Video* 7-7	☐ *Lesson Tutorial Video* 7-7
☐ Reading Strategies 7-7 CRB	☐ Problem Solving 7-7 CRB	☐ Problem Solving 7-7 CRB	☐ Reading Strategies 7-7 CRB
☐ *Questioning Strategies* pp. 118–119			☐ Lesson Vocabulary SE p. 374
☐ *IDEA Works!* 7-7			☐ *Multilingual Glossary*

ASSESSMENT
☐ Lesson Quiz, TE p. 384 and DT 7-7   ☐ State-Specific Test Prep Online Keyword: MR7 TestPrep

Copyright © Holt, Rinehart and Winston.
All rights reserved.

Holt Mathematics

Teacher's Name _____ Class _____ Date _____

Lesson Plan 7-8
Percents, Decimals, and Fractions pp. 385–388   Day _____

Objective Students write decimals and fractions as percents.

> **NCTM Standards:** Understand numbers, ways of representing numbers, relationships among numbers, and number systems; Compute fluently and make reasonable estimates; Select, apply, and translate among mathematical representations to solve problems.

Pacing
☐ 45-minute Classes: 1 day   ☐ 90-minute Classes: 1/2 day   ☐ Other_____

WARM UP
☐ Warm Up TE p. 385 and Daily Transparency 7-8
☐ Problem of the Day TE p. 385 and Daily Transparency 7-8
☐ Countdown to Testing Transparency Week 16

TEACH
☐ Lesson Presentation CD-ROM 7-8
☐ Alternate Opener, Explorations Transparency 7-8, TE p. 385, and Exploration 7-8
☐ Reaching All Learners TE p. 386
☐ Teaching Transparency 7-8
☐ *Know-It Notebook* 7-8

PRACTICE AND APPLY
☐ Example 1: Average: 1–4, 67–75 Advanced: 11–14, 65, 67–75
☐ Example 2: Average: 1–9, 62, 67–75 Advanced: 11–25, 67–75
☐ Example 3: Average: 1–18, 28–75 Advanced: 11–75

REACHING ALL LEARNERS – Differentiated Instruction for students with

Developing Knowledge	On-level Knowledge	Advanced Knowledge	English Language Development
☐ Cooperative Learning TE p. 386	☐ Cooperative Learning TE p. 386	☐ Cooperative Learning TE p. 386	☐ Cooperative Learning TE p. 386
☐ Practice A 7-8 CRB	☐ Practice B 7-8 CRB	☐ Practice C 7-8 CRB	☐ Practice A, B, or C 7-8 CRB
☐ Reteach 7-8 CRB	☐ Puzzles, Twisters & Teasers 7-8 CRB	☐ Challenge 7-8 CRB	☐ *Success for ELL* 7-8
☐ Homework Help Online Keyword: MR7 7-8	☐ Homework Help Online Keyword: MR7 7-8	☐ Homework Help Online Keyword: MR7 7-8	☐ Homework Help Online Keyword: MR7 7-8
☐ *Lesson Tutorial Video* 7-8	☐ *Lesson Tutorial Video* 7-8	☐ *Lesson Tutorial Video* 7-8	☐ *Lesson Tutorial Video* 7-8
☐ Reading Strategies 7-8 CRB	☐ Problem Solving 7-8 CRB	☐ Problem Solving 7-8 CRB	☐ Reading Strategies 7-8 CRB
☐ *Questioning Strategies* pp. 120–121			
☐ *IDEA Works!* 7-8			☐ *Multilingual Glossary*

ASSESSMENT
☐ Lesson Quiz, TE p. 388 and DT 7-8   ☐ State-Specific Test Prep Online Keyword: MR7 TestPrep

Teacher's Name _____ Class _____ Date _____

Lesson Plan 7-9
Percent Problems pp. 390–393   Day _____

Objective Students find the missing value in a percent problem.

> **NCTM Standards:** Understand numbers, ways of representing numbers, relationships among numbers, and number systems; Compute fluently and make reasonable estimates; Recognize and apply mathematics in contexts outside of mathematics.

Pacing
☐ 45-minute Classes: 1 day   ☐ 90-minute Classes: 1/2 day   ☐ Other_____

WARM UP
☐ Warm Up TE p. 390 and Daily Transparency 7-9
☐ Problem of the Day TE p. 390 and Daily Transparency 7-9
☐ Countdown to Testing Transparency Week 16

TEACH
☐ Lesson Presentation CD-ROM 7-9
☐ Alternate Opener, Explorations Transparency 7-9, TE p. 390, and Exploration 7-9
☐ Reaching All Learners TE p. 391
☐ *Technology Lab Activities* 7-9
☐ *Know-It Notebook* 7-9

PRACTICE AND APPLY
☐ Example 1: Average: 1, 19–30, 38–45 Advanced: 9–10, 19–29, 33, 38–45
☐ Example 2: Average: 1–2, 19–30, 38–45 Advanced: 9–12, 19–28, 38–45
☐ Example 3: Average: 1–26, 33–35, 38–45 Advanced: 9–45

REACHING ALL LEARNERS – Differentiated Instruction for students with

Developing Knowledge	On-level Knowledge	Advanced Knowledge	English Language Development
☐ Diversity TE p. 391	☐ Diversity TE p. 391	☐ Diversity TE p. 391	☐ Diversity TE p. 391
☐ Practice A 7-9 CRB	☐ Practice B 7-9 CRB	☐ Practice C 7-9 CRB	☐ Practice A, B, or C 7-9 CRB
☐ Reteach 7-9 CRB	☐ Puzzles, Twisters & Teasers 7-9 CRB	☐ Challenge 7-9 CRB	☐ *Success for ELL* 7-9
☐ Homework Help Online Keyword: MR7 7-9	☐ Homework Help Online Keyword: MR7 7-9	☐ Homework Help Online Keyword: MR7 7-9	☐ Homework Help Online Keyword: MR7 7-9
☐ *Lesson Tutorial Video* 7-9	☐ *Lesson Tutorial Video* 7-9	☐ *Lesson Tutorial Video* 7-9	☐ *Lesson Tutorial Video* 7-9
☐ Reading Strategies 7-9 CRB	☐ Problem Solving 7-9 CRB	☐ Problem Solving 7-9 CRB	☐ Reading Strategies 7-9 CRB
☐ *Questioning Strategies* pp. 122–123	☐ Math Connections TE p. 391	☐ Math Connections TE p. 391	
☐ *IDEA Works!* 7-9			☐ *Multilingual Glossary*

ASSESSMENT
☐ Lesson Quiz, TE p. 393 and DT 7-9   ☐ State-Specific Test Prep Online Keyword: MR7 TestPrep

Teacher's Name _____ Class _____ Date _____

Lesson Plan 7-10
Using Percents pp. 394–397   Day _____

Objective Students solve percent problems that involve discounts, tips, and sales tax.

> **NCTM Standards:** Understand numbers, ways of representing numbers, relationships among numbers, and number systems.

Pacing
- [ ] 45-minute Classes: 1 day   [ ] 90-minute Classes: 1/2 day   [ ] Other_____

WARM UP
- [ ] Warm Up TE p. 394 and Daily Transparency 7-10
- [ ] Problem of the Day TE p. 394 and Daily Transparency 7-10
- [ ] Countdown to Testing Transparency Week 16

TEACH
- [ ] Lesson Presentation CD-ROM 7-10
- [ ] Alternate Opener, Explorations Transparency 7-10, TE p. 394, and Exploration 7-10
- [ ] Reaching All Learners TE p. 395
- [ ] Teaching Transparency 7-10
- [ ] *Technology Lab Activities* 7-10
- [ ] *Know-It Notebook* 7-10

PRACTICE AND APPLY
- [ ] Example 1: Average: 1, 18–27 Advanced: 4, 15, 18–27
- [ ] Example 2: Average: 1–2, 18–27 Advanced: 4–7, 18–27
- [ ] Example 3: Average: 1–12, 16–27 Advanced: 4–27

REACHING ALL LEARNERS – Differentiated Instruction for students with

Developing Knowledge	On-level Knowledge	Advanced Knowledge	English Language Development
[ ] Home Connection TE p. 395	[ ] Home Connection TE p. 395	[ ] Home Connection TE p. 395	[ ] Home Connection TE p. 395
[ ] Practice A 7-10 CRB	[ ] Practice B 7-10 CRB	[ ] Practice C 7-10 CRB	[ ] Practice A, B, or C 7-10 CRB
[ ] Reteach 7-10 CRB	[ ] Puzzles, Twisters & Teasers 7-10 CRB	[ ] Challenge 7-10 CRB	[ ] *Success for ELL* 7-10
[ ] Homework Help Online Keyword: MR7 7-10	[ ] Homework Help Online Keyword: MR7 7-10	[ ] Homework Help Online Keyword: MR7 7-10	[ ] Homework Help Online Keyword: MR7 7-10
[ ] *Lesson Tutorial Video* 7-10	[ ] *Lesson Tutorial Video* 7-10	[ ] *Lesson Tutorial Video* 7-10	[ ] *Lesson Tutorial Video* 7-10
[ ] Reading Strategies 7-10 CRB	[ ] Problem Solving 7-10 CRB	[ ] Problem Solving 7-10 CRB	[ ] Reading Strategies 7-10 CRB
[ ] *Questioning Strategies* pp. 124–125	[ ] Graphic Organizers TE p. 395	[ ] Graphic Organizers TE p. 395	[ ] Lesson Vocabulary SE p. 394
[ ] *IDEA Works!* 7-10			[ ] *Multilingual Glossary*

ASSESSMENT
- [ ] Lesson Quiz, TE p. 397 and DT 7-10   [ ] State-Specific Test Prep Online Keyword: MR7 TestPrep

Teacher's Name _____ Class _____ Date _____

Lesson Plan 8-1
Building Blocks of Geometry pp. 416–419 Day _____

Objective Students describe figures by using the terms of geometry.

> **NCTM Standards:** Analyze characteristics and properties of two- and three-dimensional geometric shapes and develop mathematical arguments about geometric relationships; Use visualization, spatial reasoning, and geometric modeling to solve problems.

Pacing
☐ 45-minute Classes: 1 day ☐ 90-minute Classes: 1/2 day ☐ Other_____

WARM UP
☐ Warm Up TE p. 416 and Daily Transparency 8-1
☐ Problem of the Day TE p. 416 and Daily Transparency 8-1
☐ Countdown to Testing Transparency Week 17

TEACH
☐ Lesson Presentation CD-ROM 8-1
☐ Alternate Opener, Explorations Transparency 8-1, TE p. 416, and Exploration 8-1
☐ Reaching All Learners TE p. 417
☐ Teaching Transparency 8-1
☐ *Know-It Notebook* 8-1

PRACTICE AND APPLY
☐ Example 1: Average: 1–4, 15–19, 31–41 Advanced: 8–11, 15–19, 31–41
☐ Example 2: Average: 1–14, 20–41 Advanced: 8–41

REACHING ALL LEARNERS – Differentiated Instruction for students with

Developing Knowledge	On-level Knowledge	Advanced Knowledge	English Language Development
☐ Inclusion TE p. 417	☐ Cognitive Strategies TE p. 417	☐ Cognitive Strategies TE p. 417	☐ Cognitive Strategies TE p. 417
☐ Practice A 8-1 CRB	☐ Practice B 8-1 CRB	☐ Practice C 8-1 CRB	☐ Practice A, B, or C 8-1 CRB
☐ Reteach 8-1 CRB	☐ Puzzles, Twisters & Teasers 8-1 CRB	☐ Challenge 8-1 CRB	☐ *Success for ELL* 8-1
☐ Homework Help Online Keyword: MR7 8-1	☐ Homework Help Online Keyword: MR7 8-1	☐ Homework Help Online Keyword: MR7 8-1	☐ Homework Help Online Keyword: MR7 8-1
☐ *Lesson Tutorial Video* 8-1	☐ *Lesson Tutorial Video* 8-1	☐ *Lesson Tutorial Video* 8-1	☐ *Lesson Tutorial Video* 8-1
☐ Reading Strategies 8-1 CRB	☐ Problem Solving 8-1 CRB	☐ Problem Solving 8-1 CRB	☐ Reading Strategies 8-1 CRB
☐ *Questioning Strategies* pp. 126–127			☐ *Lesson Vocabulary* SE p. 416
☐ *IDEA Works!* 8-1			☐ *Multilingual Glossary*

ASSESSMENT
☐ Lesson Quiz, TE p. 419 and DT 8-1 ☐ State-Specific Test Prep Online Keyword: MR7 TestPrep

Copyright © Holt, Rinehart and Winston.
All rights reserved.

Holt Mathematics

Teacher's Name _____ Class _____ Date _____

Lesson Plan 8-2
Measuring and Classifying Angles pp. 420–423   Day _____

Objective Students learn to name, measure, draw, and classify angles.

> **NCTM Standards:** Analyze characteristics and properties of two- and three-dimensional geometric shapes and develop mathematical arguments about geometric relationships; Understand measurable attributes of objects and the units, systems, and processes of measurement.

Pacing
☐ 45-minute Classes: 1 day   ☐ 90-minute Classes: 1/2 day   ☐ Other _____

WARM UP
☐ Warm Up TE p. 420 and Daily Transparency 8-2
☐ Problem of the Day TE p. 420 and Daily Transparency 8-2
☐ Countdown to Testing Transparency Week 17

TEACH
☐ Lesson Presentation CD-ROM 8-2
☐ Alternate Opener, Explorations Transparency 8-2, TE p. 420, and Exploration 8-2
☐ Reaching All Learners TE p. 421
☐ Teaching Transparency 8-2
☐ *Know-It Notebook* 8-2

PRACTICE AND APPLY
☐ Example 1: Average: 1–3, 34–44   Advanced: 12–14, 34–44
☐ Example 2: Average: 1–7, 24–25, 34–44   Advanced: 12–19, 26, 34–44
☐ Example 3: Average: 1–10, 24–29, 34–44   Advanced: 12–22, 24–28, 34–44
☐ Example 3: Average: 1–11, 15–23, 30–44   Advanced: 12–44

REACHING ALL LEARNERS – Differentiated Instruction for students with

Developing Knowledge	On-level Knowledge	Advanced Knowledge	English Language Development
☐ Visual Cues TE p. 421	☐ Visual Cues TE p. 421	☐ Visual Cues TE p. 421	☐ Visual Cues TE p. 421
☐ Practice A 8-2 CRB	☐ Practice B 8-2 CRB	☐ Practice C 8-2 CRB	☐ Practice A, B, or C 8-2 CRB
☐ Reteach 8-2 CRB	☐ Puzzles, Twisters & Teasers 8-2 CRB	☐ Challenge 8-2 CRB	☐ *Success for ELL* 8-2
☐ Homework Help Online Keyword: MR7 8-2	☐ Homework Help Online Keyword: MR7 8-2	☐ Homework Help Online Keyword: MR7 8-2	☐ Homework Help Online Keyword: MR7 8-2
☐ *Lesson Tutorial Video* 8-2	☐ *Lesson Tutorial Video* 8-2	☐ *Lesson Tutorial Video* 8-2	☐ *Lesson Tutorial Video* 8-2
☐ Reading Strategies 8-2 CRB	☐ Problem Solving 8-2 CRB	☐ Problem Solving 8-2 CRB	☐ Reading Strategies 8-2 CRB
☐ *Questioning Strategies* pp. 128–129	☐ Language Arts TE p. 421	☐ Language Arts TE p. 421	☐ Lesson Vocabulary SE p. 420
☐ *IDEA Works!* 8-2			☐ *Multilingual Glossary*

ASSESSMENT
☐ Lesson Quiz, TE p. 423 and DT 8-2   ☐ State-Specific Test Prep Online Keyword: MR7 TestPrep

Teacher's Name _____ Class _____ Date _____

Lesson Plan 8-3
Angle Relationships pp. 424–427   Day _____

Objective Students understand relationships of angles.

> **NCTM Standards:** Analyze characteristics and properties of two- and three-dimensional geometric shapes and develop mathematical arguments about geometric relationships.

Pacing
- ☐ 45-minute Classes: 1 day   ☐ 90-minute Classes: 1/2 day   ☐ Other_____

WARM UP
- ☐ Warm Up TE p. 424 and Daily Transparency 8-3
- ☐ Problem of the Day TE p. 424 and Daily Transparency 8-3
- ☐ Countdown to Testing Transparency Week 17

TEACH
- ☐ Lesson Presentation CD-ROM 8-3
- ☐ Alternate Opener, Explorations Transparency 8-3, TE p. 424, and Exploration 8-3
- ☐ Reaching All Learners TE p. 425
- ☐ Teaching Transparency 8-3
- ☐ *Hands-On Lab Activities* 8-3
- ☐ *Technology Lab Activities* 8-3
- ☐ *Know-It Notebook* 8-3

PRACTICE AND APPLY
- ☐ Example 1: Average: 1–2, 29–38 Advanced: 5–6, 29–38
- ☐ Example 2: Average: 1–4, 9–38 Advanced: 5–38

REACHING ALL LEARNERS – Differentiated Instruction for students with

Developing Knowledge	On-level Knowledge	Advanced Knowledge	English Language Development
☐ Inclusion TE p. 425	☐ Concrete Manipulatives TE p. 425	☐ Concrete Manipulatives TE p. 425	☐ Concrete Manipulatives TE p. 425
☐ Practice A 8-3 CRB	☐ Practice B 8-3 CRB	☐ Practice C 8-3 CRB	☐ Practice A, B, or C 8-3 CRB
☐ Reteach 8-3 CRB	☐ Puzzles, Twisters & Teasers 8-3 CRB	☐ Challenge 8-3 CRB	☐ *Success for ELL* 8-3
☐ Homework Help Online Keyword: MR7 8-3	☐ Homework Help Online Keyword: MR7 8-3	☐ Homework Help Online Keyword: MR7 8-3	☐ Homework Help Online Keyword: MR7 8-3
☐ *Lesson Tutorial Video* 8-3	☐ *Lesson Tutorial Video* 8-3	☐ *Lesson Tutorial Video* 8-3	☐ *Lesson Tutorial Video* 8-3
☐ Reading Strategies 8-3 CRB	☐ Problem Solving 8-3 CRB	☐ Problem Solving 8-3 CRB	☐ Reading Strategies 8-3 CRB
☐ *Questioning Strategies* pp. 130–131			☐ Lesson Vocabulary SE p. 424
☐ *IDEA Works!* 8-3			☐ Multilingual Glossary

ASSESSMENT
- ☐ Lesson Quiz, TE p. 427 and DT 8-3   ☐ State-Specific Test Prep Online Keyword: MR7 TestPrep

Teacher's Name _____ Class _____ Date _____

Lesson Plan 8-4
Classifying Lines pp. 428–431   Day _____

Objective Students classify the different types of lines.

> **NCTM Standards:** Analyze characteristics and properties of two- and three-dimensional geometric shapes and develop mathematical arguments about geometric relationships.

Pacing
☐ 45-minute Classes: 1 day   ☐ 90-minute Classes: 1/2 day   ☐ Other_____

WARM UP
☐ Warm Up TE p. 428 and Daily Transparency 8-4
☐ Problem of the Day TE p. 428 and Daily Transparency 8-4
☐ Countdown to Testing Transparency Week 17

TEACH
☐ Lesson Presentation CD-ROM 8-4
☐ Alternate Opener, Explorations Transparency 8-4, TE p. 428, and Exploration 8-4
☐ Reaching All Learners TE p. 429
☐ Teaching Transparency 8-4
☐ *Hands-On Lab Activities* 8-4
☐ *Technology Lab Activities* 8-4
☐ *Know-It Notebook* 8-4

PRACTICE AND APPLY
☐ Example 1: Average: 1–2, 8–10, 24–36 Advanced: 4–6, 8–10, 24–36
☐ Example 2: Average: 1–9, 12–36 Advanced: 4–36

REACHING ALL LEARNERS – Differentiated Instruction for students with

Developing Knowledge	On-level Knowledge	Advanced Knowledge	English Language Development
☐ Cognitive Strategies TE p. 429	☐ Cognitive Strategies TE p. 429	☐ Cognitive Strategies TE p. 429	☐ Cognitive Strategies TE p. 429
☐ Practice A 8-4 CRB	☐ Practice B 8-4 CRB	☐ Practice C 8-4 CRB	☐ Practice A, B, or C 8-4 CRB
☐ Reteach 8-4 CRB	☐ Puzzles, Twisters & Teasers 8-4 CRB	☐ Challenge 8-4 CRB	☐ *Success for ELL* 8-4
☐ Homework Help Online Keyword: MR7 8-4	☐ Homework Help Online Keyword: MR7 8-4	☐ Homework Help Online Keyword: MR7 8-4	☐ Homework Help Online Keyword: MR7 8-4
☐ *Lesson Tutorial Video* 8-4	☐ *Lesson Tutorial Video* 8-4	☐ *Lesson Tutorial Video* 8-4	☐ *Lesson Tutorial Video* 8-4
☐ Reading Strategies 8-4 CRB	☐ Problem Solving 8-4 CRB	☐ Problem Solving 8-4 CRB	☐ Reading Strategies 8-4 CRB
☐ *Questioning Strategies* pp. 132–133	☐ Multiple Representations TE p. 429	☐ Multiple Representations TE p. 429	☐ Lesson Vocabulary SE p. 428
☐ *IDEA Works!* 8-4			☐ *Multilingual Glossary*

ASSESSMENT
☐ Lesson Quiz, TE p. 431 and DT 8-4   ☐ State-Specific Test Prep Online Keyword: MR7 TestPrep

Holt Mathematics

Teacher's Name _____ Class _____ Date _____

Lesson Plan 8-5
Triangles pp. 437–440   Day _____

Objective Students classify triangles and solve problems involving angle and side measures of triangles.

> **NCTM Standards:** Analyze characteristics and properties of two- and three-dimensional geometric shapes and develop mathematical arguments about geometric relationships; Use visualization, spatial reasoning, and geometric modeling to solve problems; Develop and evaluate mathematical arguments and proofs.

Pacing
☐ 45-minute Classes: 1 day   ☐ 90-minute Classes: 1/2 day   ☐ Other _____

WARM UP
☐ Warm Up TE p. 437 and Daily Transparency 8-5
☐ Problem of the Day TE p. 437 and Daily Transparency 8-5
☐ Countdown to Testing Transparency Week 18

TEACH
☐ Lesson Presentation CD-ROM 8-5
☐ Alternate Opener, Explorations Transparency 8-5, TE p. 437, and Exploration 8-5
☐ Reaching All Learners TE p. 438
☐ Teaching Transparency 8-5
☐ *Know-It Notebook* 8-5

PRACTICE AND APPLY
☐ Example 1: Average: 1, 11–16, 29–39 Advanced: 6, 11–16, 29–39
☐ Example 2: Average: 1–3, 11–16, 29–39 Advanced: 6–8, 11–16, 29–39
☐ Example 3: Average: 1–13, 18–39 Advanced: 6–39

REACHING ALL LEARNERS – Differentiated Instruction for students with

Developing Knowledge	On-level Knowledge	Advanced Knowledge	English Language Development
☐ Inclusion TE p. 438	☐ Modeling TE p. 438	☐ Modeling TE p. 438	☐ Modeling TE p. 438
☐ Practice A 8-5 CRB	☐ Practice B 8-5 CRB	☐ Practice C 8-5 CRB	☐ Practice A, B, or C 8-5 CRB
☐ Reteach 8-5 CRB	☐ Puzzles, Twisters & Teasers 8-5 CRB	☐ Challenge 8-5 CRB	☐ *Success for ELL* 8-5
☐ Homework Help Online Keyword: MR7 8-5	☐ Homework Help Online Keyword: MR7 8-5	☐ Homework Help Online Keyword: MR7 8-5	☐ Homework Help Online Keyword: MR7 8-5
☐ *Lesson Tutorial Video* 8-5	☐ *Lesson Tutorial Video* 8-5	☐ *Lesson Tutorial Video* 8-5	☐ *Lesson Tutorial Video* 8-5
☐ Reading Strategies 8-5 CRB	☐ Problem Solving 8-5 CRB	☐ Problem Solving 8-5 CRB	☐ Reading Strategies 8-5 CRB
☐ *Questioning Strategies* pp. 134–135			☐ Lesson Vocabulary SE p. 437
☐ *IDEA Works!* 8-5			☐ *Multilingual Glossary*

ASSESSMENT
☐ Lesson Quiz, TE p. 440 and DT 8-5   ☐ State-Specific Test Prep Online Keyword: MR7 TestPrep

Teacher's Name _____ Class _____ Date _____

Lesson Plan 8-6
Quadrilaterals pp. 442–445   Day _____

Objective Students identify, classify, and compare quadrilaterals.

> **NCTM Standards:** Analyze characteristics and properties of two- and three-dimensional geometric shapes and develop mathematical arguments about geometric relationships.

Pacing
☐ 45-minute Classes: 1 day   ☐ 90-minute Classes: 1/2 day   ☐ Other_____

WARM UP
☐ Warm Up TE p. 442 and Daily Transparency 8-6
☐ Problem of the Day TE p. 442 and Daily Transparency 8-6
☐ Countdown to Testing Transparency Week 18

TEACH
☐ Lesson Presentation CD-ROM 8-6
☐ Alternate Opener, Explorations Transparency 8-6, TE p. 442, and Exploration 8-6
☐ Reaching All Learners TE p. 443
☐ Teaching Transparency 8-6
☐ *Know-It Notebook* 8-6

PRACTICE AND APPLY
☐ Example 1: Average: 1–3, 13–15, 34–39 Advanced: 7–9, 13–15, 34–39
☐ Example 2: Average: 1–19, 25–39 Advanced: 7–39

REACHING ALL LEARNERS – Differentiated Instruction for students with

Developing Knowledge	On-level Knowledge	Advanced Knowledge	English Language Development
☐ Diversity TE p. 443	☐ Diversity TE p. 443	☐ Diversity TE p. 443	☐ Diversity TE p. 443
☐ Practice A 8-6 CRB	☐ Practice B 8-6 CRB	☐ Practice C 8-6 CRB	☐ Practice A, B, or C 8-6 CRB
☐ Reteach 8-6 CRB	☐ Puzzles, Twisters & Teasers 8-6 CRB	☐ Challenge 8-6 CRB	☐ *Success for ELL* 8-6
☐ Homework Help Online Keyword: MR7 8-6	☐ Homework Help Online Keyword: MR7 8-6	☐ Homework Help Online Keyword: MR7 8-6	☐ Homework Help Online Keyword: MR7 8-6
☐ *Lesson Tutorial Video* 8-6	☐ *Lesson Tutorial Video* 8-6	☐ *Lesson Tutorial Video* 8-6	☐ *Lesson Tutorial Video* 8-6
☐ Reading Strategies 8-6 CRB	☐ Problem Solving 8-6 CRB	☐ Problem Solving 8-6 CRB	☐ Reading Strategies 8-6 CRB
☐ *Questioning Strategies* pp. 136–137	☐ Kinesthetic Experience TE p. 443	☐ Kinesthetic Experience TE p. 443	☐ Lesson Vocabulary SE p. 442
☐ *IDEA Works!* 8-6			☐ *Multilingual Glossary*

ASSESSMENT
☐ Lesson Quiz, TE p. 445 and DT 8-6   ☐ State-Specific Test Prep Online Keyword: MR7 TestPrep

Holt Mathematics

Teacher's Name _____ Class _____ Date _____

Lesson Plan 8-7
Polygons pp. 446–449   Day _____

Objective Students identify regular and not regular polygons and find the angle measures of regular polygons.

> **NCTM Standards:** Analyze characteristics and properties of two- and three-dimensional geometric shapes and develop mathematical arguments about geometric relationships; Use visualization, spatial reasoning, and geometric modeling to solve problems; Recognize reasoning and proof as fundamental aspects of mathematics.

Pacing
☐ 45-minute Classes: 1 day   ☐ 90-minute Classes: 1/2 day   ☐ Other_____

WARM UP
☐ Warm Up TE p. 446 and Daily Transparency 8-7
☐ Problem of the Day TE p. 446 and Daily Transparency 8-7
☐ Countdown to Testing Transparency Week 18

TEACH
☐ Lesson Presentation CD-ROM 8-7
☐ Alternate Opener, Explorations Transparency 8-7, TE p. 446, and Exploration 8-7
☐ Reaching All Learners TE p. 447
☐ Teaching Transparency 8-7
☐ *Hands-On Lab Activities* 8-7
☐ *Know-It Notebook* 8-7

PRACTICE AND APPLY
☐ Example 1: Average: 1–3, 9–14, 24–30 Advanced: 5–7, 9–14, 24–30
☐ Example 2: Average: 1–4, 9–30 Advanced: 5–30

REACHING ALL LEARNERS – Differentiated Instruction for students with

Developing Knowledge	On-level Knowledge	Advanced Knowledge	English Language Development
☐ Critical Thinking TE p. 447	☐ Critical Thinking TE p. 447	☐ Critical Thinking TE p. 447	☐ Critical Thinking TE p. 447
☐ Practice A 8-7 CRB	☐ Practice B 8-7 CRB	☐ Practice C 8-7 CRB	☐ Practice A, B, or C 8-7 CRB
☐ Reteach 8-7 CRB	☐ Puzzles, Twisters & Teasers 8-7 CRB	☐ Challenge 8-7 CRB	☐ *Success for ELL* 8-7
☐ Homework Help Online Keyword: MR7 8-7	☐ Homework Help Online Keyword: MR7 8-7	☐ Homework Help Online Keyword: MR7 8-7	☐ Homework Help Online Keyword: MR7 8-7
☐ *Lesson Tutorial Video* 8-7	☐ *Lesson Tutorial Video* 8-7	☐ *Lesson Tutorial Video* 8-7	☐ *Lesson Tutorial Video* 8-7
☐ Reading Strategies 8-7 CRB	☐ Problem Solving 8-7 CRB	☐ Problem Solving 8-7 CRB	☐ Reading Strategies 8-7 CRB
☐ *Questioning Strategies* pp. 138–139	☐ Language Arts TE p. 447	☐ Language Arts TE p. 447	☐ Lesson Vocabulary SE p. 446
☐ *IDEA Works!* 8-7			☐ *Multilingual Glossary*

ASSESSMENT
☐ Lesson Quiz, TE p. 449 and DT 8-7   ☐ State-Specific Test Prep Online Keyword: MR7 TestPrep

Teacher's Name _____ Class _____ Date _____

Lesson Plan 8-8
Geometric Patterns pp. 450–453 Day _____

Objective Students recognize, describe, and extend geometric patterns.

> **NCTM Standards:** Use visualization, spatial reasoning, and geometric modeling to solve problems.

Pacing
☐ 45-minute Classes: 1 day ☐ 90-minute Classes: 1/2 day ☐ Other_____

WARM UP
☐ Warm Up TE p. 450 and Daily Transparency 8-8
☐ Problem of the Day TE p. 450 and Daily Transparency 8-8
☐ Countdown to Testing Transparency Week 19

TEACH
☐ Lesson Presentation CD-ROM 8-8
☐ Alternate Opener, Explorations Transparency 8-8, TE p. 450, and Exploration 8-8
☐ Reaching All Learners TE p. 451
☐ Teaching Transparency 8-8
☐ *Hands-On Lab Activities* 8-8
☐ *Know-It Notebook* 8-8

PRACTICE AND APPLY
☐ Example 1: Average: 1, 7–9, 13–24 Advanced: 4, 7–9, 13–24
☐ Example 2: Average: 1–2, 7–9, 13–24 Advanced: 4–5, 7–9, 13–24
☐ Example 3: Average: 1–6, 10–24 Advanced: 4–24

REACHING ALL LEARNERS – Differentiated Instruction for students with

Developing Knowledge	On-level Knowledge	Advanced Knowledge	English Language Development
☐ Modeling TE p. 451	☐ Modeling TE p. 451	☐ Modeling TE p. 451	☐ Modeling TE p. 451
☐ Practice A 8-8 CRB	☐ Practice B 8-8 CRB	☐ Practice C 8-8 CRB	☐ Practice A, B, or C 8-8 CRB
☐ Reteach 8-8 CRB	☐ Puzzles, Twisters & Teasers 8-8 CRB	☐ Challenge 8-8 CRB	☐ *Success for ELL* 8-8
☐ Homework Help Online Keyword: MR7 8-8	☐ Homework Help Online Keyword: MR7 8-8	☐ Homework Help Online Keyword: MR7 8-8	☐ Homework Help Online Keyword: MR7 8-8
☐ *Lesson Tutorial Video* 8-8	☐ *Lesson Tutorial Video* 8-8	☐ *Lesson Tutorial Video* 8-8	☐ *Lesson Tutorial Video* 8-8
☐ Reading Strategies 8-8 CRB	☐ Problem Solving 8-8 CRB	☐ Problem Solving 8-8 CRB	☐ Reading Strategies 8-8 CRB
☐ *Questioning Strategies* pp. 140–141	☐ Multiple Representations TE p. 451	☐ Multiple Representations TE p. 451	
☐ *IDEA Works!* 8-8			☐ *Multilingual Glossary*

ASSESSMENT
☐ Lesson Quiz, TE p. 453 and DT 8-8 ☐ State-Specific Test Prep Online Keyword: MR7 TestPrep

Teacher's Name _____ Class _____ Date _____

Lesson Plan 8-9
Congruence pp. 456–458   Day _____

Objective Students identify congruent figures and use congruence to solve problems.

> **NCTM Standards:** Analyze characteristics and properties of two- and three-dimensional geometric shapes and develop mathematical arguments about geometric relationships.

Pacing
- [ ] 45-minute Classes: 1 day   [ ] 90-minute Classes: 1/2 day   [ ] Other_____

WARM UP
- [ ] Warm Up TE p. 456 and Daily Transparency 8-9
- [ ] Problem of the Day TE p. 456 and Daily Transparency 8-9
- [ ] Countdown to Testing Transparency Week 19

TEACH
- [ ] Lesson Presentation CD-ROM 8-9
- [ ] Alternate Opener, Explorations Transparency 8-9, TE p. 456, and Exploration 8-9
- [ ] Reaching All Learners TE p. 457
- [ ] *Know-It Notebook* 8-9

PRACTICE AND APPLY
- [ ] Example 1: Average: 1–2, 11–20 Advanced: 4–5, 11–20
- [ ] Example 2: Average: 1–6, 8–20 Advanced: 4–20

REACHING ALL LEARNERS – Differentiated Instruction for students with

Developing Knowledge	On-level Knowledge	Advanced Knowledge	English Language Development
[ ] Cooperative Learning TE p. 457	[ ] Cooperative Learning TE p. 457	[ ] Cooperative Learning TE p. 457	[ ] Cooperative Learning TE p. 457
[ ] Practice A 8-9 CRB	[ ] Practice B 8-9 CRB	[ ] Practice C 8-9 CRB	[ ] Practice A, B, or C 8-9 CRB
[ ] Reteach 8-9 CRB	[ ] Puzzles, Twisters & Teasers 8-9 CRB	[ ] Challenge 8-9 CRB	[ ] *Success for ELL* 8-9
[ ] Homework Help Online Keyword: MR7 8-9	[ ] Homework Help Online Keyword: MR7 8-9	[ ] Homework Help Online Keyword: MR7 8-9	[ ] Homework Help Online Keyword: MR7 8-9
[ ] *Lesson Tutorial Video* 8-9	[ ] *Lesson Tutorial Video* 8-9	[ ] *Lesson Tutorial Video* 8-9	[ ] *Lesson Tutorial Video* 8-9
[ ] Reading Strategies 8-9 CRB	[ ] Problem Solving 8-9 CRB	[ ] Problem Solving 8-9 CRB	[ ] Reading Strategies 8-9 CRB
[ ] *Questioning Strategies* pp. 142–143			
[ ] *IDEA Works!* 8-9			[ ] *Multilingual Glossary*

ASSESSMENT
- [ ] Lesson Quiz, TE p. 458 and DT 8-9   [ ] State-Specific Test Prep Online Keyword: MR7 TestPrep

Teacher's Name _____ Class _____ Date _____

Lesson Plan 8-10
Transformations pp. 459–462 Day _____

Objective Students use translations, reflections, and rotations to transform geometric shapes.

> **NCTM Standards:** Apply transformations and use symmetry to analyze mathematical situations; Use visualization, spatial reasoning, and geometric modeling to solve problems.

Pacing
- ☐ 45-minute Classes: 1 day ☐ 90-minute Classes: 1/2 day ☐ Other_____

WARM UP
- ☐ Warm Up TE p. 459 and Daily Transparency 8-10
- ☐ Problem of the Day TE p. 459 and Daily Transparency 8-10
- ☐ Countdown to Testing Transparency Week 19

TEACH
- ☐ Lesson Presentation CD-ROM 8-10
- ☐ Alternate Opener, Explorations Transparency 8-10, TE p. 459, and Exploration 8-10
- ☐ Reaching All Learners TE p. 460
- ☐ Teaching Transparency 8-10
- ☐ *Hands-On Lab Activities* 8-10
- ☐ *Technology Lab Activities* 8-10
- ☐ *Know-It Notebook* 8-10

PRACTICE AND APPLY
- ☐ Example 1: Average: 1–3, 15, 21–30 Advanced: 6–8, 15, 21–30
- ☐ Example 2: Average: 1–11, 16–30 Advanced: 6–30

REACHING ALL LEARNERS – Differentiated Instruction for students with

Developing Knowledge	On-level Knowledge	Advanced Knowledge	English Language Development
☐ Concrete Manipulatives TE p. 460	☐ Concrete Manipulatives TE p. 460	☐ Concrete Manipulatives TE p. 460	☐ Concrete Manipulatives TE p. 460
☐ Practice A 8-10 CRB	☐ Practice B 8-10 CRB	☐ Practice C 8-10 CRB	☐ Practice A, B, or C 8-10 CRB
☐ Reteach 8-10 CRB	☐ Puzzles, Twisters & Teasers 8-10 CRB	☐ Challenge 8-10 CRB	☐ *Success for ELL* 8-10
☐ Homework Help Online Keyword: MR7 8-10	☐ Homework Help Online Keyword: MR7 8-10	☐ Homework Help Online Keyword: MR7 8-10	☐ Homework Help Online Keyword: MR7 8-10
☐ *Lesson Tutorial Video* 8-10	☐ *Lesson Tutorial Video* 8-10	☐ *Lesson Tutorial Video* 8-10	☐ *Lesson Tutorial Video* 8-10
☐ Reading Strategies 8-10 CRB	☐ Problem Solving 8-10 CRB	☐ Problem Solving 8-10 CRB	☐ Reading Strategies 8-10 CRB
☐ *Questioning Strategies* pp. 144–145	☐ Kinesthetic Experience TE p. 460	☐ Kinesthetic Experience TE p. 460	☐ Lesson Vocabulary SE p. 459
☐ *IDEA Works!* 8-10			☐ *Multilingual Glossary*

ASSESSMENT
- ☐ Lesson Quiz, TE p. 462 and DT 8-10   ☐ State-Specific Test Prep Online Keyword: MR7 TestPrep

Copyright © Holt, Rinehart and Winston.
All rights reserved.

Holt Mathematics

Teacher's Name _____ Class _____ Date _____

Lesson Plan 8-11
Line Symmetry pp. 464–467   Day _____

Objective Students identify line symmetry.

> **NCTM Standards:** Apply transformations and use symmetry to analyze mathematical situations; Use visualization, spatial reasoning, and geometric modeling to solve problems.

Pacing
☐ 45-minute Classes: 1 day   ☐ 90-minute Classes: 1/2 day   ☐ Other_____

WARM UP
☐ Warm Up TE p. 464 and Daily Transparency 8-11
☐ Problem of the Day TE p. 464 and Daily Transparency 8-11
☐ Countdown to Testing Transparency Week 19

TEACH
☐ Lesson Presentation CD-ROM 8-11
☐ Alternate Opener, Explorations Transparency 8-11, TE p. 464, and Exploration 8-11
☐ Reaching All Learners TE p. 465
☐ Teaching Transparency 8-11
☐ *Hands-On Lab Activities* 8-11
☐ *Know-It Notebook* 8-11

PRACTICE AND APPLY
☐ Example 1: Average: 1–3, 18, 21–28 Advanced: 9–11, 18, 21–28
☐ Example 2: Average: 1–6, 18, 21–28 Advanced: 9–14, 18, 21–28
☐ Example 3: Average: 1–17, 21–28 Advanced: 9–28

REACHING ALL LEARNERS – Differentiated Instruction for students with

Developing Knowledge	On-level Knowledge	Advanced Knowledge	English Language Development
☐ Curriculum Integration TE p. 465	☐ Curriculum Integration TE p. 465	☐ Curriculum Integration TE p. 465	☐ Curriculum Integration TE p. 465
☐ Practice A 8-11 CRB	☐ Practice B 8-11 CRB	☐ Practice C 8-11 CRB	☐ Practice A, B, or C 8-11 CRB
☐ Reteach 8-11 CRB	☐ Puzzles, Twisters & Teasers 8-11 CRB	☐ Challenge 8-11 CRB	☐ *Success for ELL* 8-11
☐ Homework Help Online Keyword: MR7 8-11	☐ Homework Help Online Keyword: MR7 8-11	☐ Homework Help Online Keyword: MR7 8-11	☐ Homework Help Online Keyword: MR7 8-11
☐ *Lesson Tutorial Video* 8-11	☐ *Lesson Tutorial Video* 8-11	☐ *Lesson Tutorial Video* 8-11	☐ *Lesson Tutorial Video* 8-11
☐ Reading Strategies 8-11 CRB	☐ Problem Solving 8-11 CRB	☐ Problem Solving 8-11 CRB	☐ Reading Strategies 8-11 CRB
☐ *Questioning Strategies* pp. 146–147			☐ Lesson Vocabulary SE p. 464
☐ *IDEA Works!* 8-11			☐ *Multilingual Glossary*

ASSESSMENT
☐ Lesson Quiz, TE p. 467 and DT 8-11   ☐ State-Specific Test Prep Online Keyword: MR7 TestPrep

Teacher's Name _____ Class _____ Date _____

Lesson Plan 9-1
Understanding Customary Units of Measure pp. 488–491   Day _____

Objective Students understand and select appropriate units of measure.

> **NCTM Standards:** Understand measurable attributes of objects and the units, systems, and processes of measurement.

Pacing
☐ 45-minute Classes: 1 day   ☐ 90-minute Classes: 1/2 day   ☐ Other _____

WARM UP
☐ Warm Up TE p. 488 and Daily Transparency 9-1
☐ Problem of the Day TE p. 488 and Daily Transparency 9-1
☐ Countdown to Testing Transparency Week 20

TEACH
☐ Lesson Presentation CD-ROM 9-1
☐ Alternate Opener, Explorations Transparency 9-1, TE p. 488, and Exploration 9-1
☐ Reaching All Learners TE p. 489
☐ Teaching Transparency 9-1
☐ *Know-It Notebook* 9-1

PRACTICE AND APPLY
☐ Example 1: Average: 1, 9–10, 15–16, 28–37 Advanced: 5, 9–10, 15, 22, 28–37
☐ Example 2: Average: 1–2, 9–10, 13–16, 18, 28–37 Advanced: 5–6, 9–10, 13–16, 22, 28–37
☐ Example 3: Average: 1–3, 9–18, 28–37 Advanced: 5–7, 9–17, 22, 28–37
☐ Example 4: Average: 1–20, 25–37 Advanced: 5–37

REACHING ALL LEARNERS – Differentiated Instruction for students with

Developing Knowledge	On-level Knowledge	Advanced Knowledge	English Language Development
☐ Diversity TE p. 489	☐ Diversity TE p. 489	☐ Diversity TE p. 489	☐ Diversity TE p. 489
☐ Practice A 9-1 CRB	☐ Practice B 9-1 CRB	☐ Practice C 9-1 CRB	☐ Practice A, B, or C 9-1 CRB
☐ Reteach 9-1 CRB	☐ Puzzles, Twisters & Teasers 9-1 CRB	☐ Challenge 9-1 CRB	☐ *Success for ELL* 9-1
☐ Homework Help Online Keyword: MR7 9-1	☐ Homework Help Online Keyword: MR7 9-1	☐ Homework Help Online Keyword: MR7 9-1	☐ Homework Help Online Keyword: MR7 9-1
☐ *Lesson Tutorial Video* 9-1	☐ *Lesson Tutorial Video* 9-1	☐ *Lesson Tutorial Video* 9-1	☐ *Lesson Tutorial Video* 9-1
☐ Reading Strategies 9-1 CRB	☐ Problem Solving 9-1 CRB	☐ Problem Solving 9-1 CRB	☐ Reading Strategies 9-1 CRB
☐ *Questioning Strategies* pp. 148–149			☐ Lesson Vocabulary SE p. 488
☐ *IDEA Works!* 9-1			☐ *Multilingual Glossary*

ASSESSMENT
☐ Lesson Quiz, TE p. 491 and DT 9-1   ☐ State-Specific Test Prep Online Keyword: MR7 TestPrep

Holt Mathematics

Teacher's Name _____ Class _____ Date _____

Lesson Plan 9-2
Understanding Metric Units of Measure pp. 492–495 Day _____

Objective Students understand and select appropriate metric units of measure.

> **NCTM Standards:** Understand measurable attributes of objects and the units, systems, and processes of measurement.

Pacing
☐ 45-minute Classes: 1 day   ☐ 90-minute Classes: 1/2 day   ☐ Other_____

WARM UP
☐ Warm Up TE p. 492 and Daily Transparency 9-2
☐ Problem of the Day TE p. 492 and Daily Transparency 9-2
☐ Countdown to Testing Transparency Week 20

TEACH
☐ Lesson Presentation CD-ROM 9-2
☐ Alternate Opener, Explorations Transparency 9-2, TE p. 492, and Exploration 9-2
☐ Reaching All Learners TE p. 493
☐ Teaching Transparency 9-2
☐ *Hands-On Lab Activities* 9-2
☐ *Know-It Notebook* 9-2

PRACTICE AND APPLY
☐ Example 1: Average: 1, 11–13, 23–32 Advanced: 6, 11–13, 23–32
☐ Example 2: Average: 1–2, 11–15, 23–32 Advanced: 6–7, 11–14, 18, 23–32
☐ Example 3: Average: 1–4, 11–18, 23–32 Advanced: 6–9, 11–17, 19, 23–32
☐ Example 4: Average: 1–12, 18–32 Advanced: 6–32

REACHING ALL LEARNERS – Differentiated Instruction for students with

Developing Knowledge	On-level Knowledge	Advanced Knowledge	English Language Development
☐ Kinesthetic Experience TE p. 493	☐ Kinesthetic Experience TE p. 493	☐ Kinesthetic Experience TE p. 493	☐ Kinesthetic Experience TE p. 493
☐ Practice A 9-2 CRB	☐ Practice B 9-2 CRB	☐ Practice C 9-2 CRB	☐ Practice A, B, or C 9-2 CRB
☐ Reteach 9-2 CRB	☐ Puzzles, Twisters & Teasers 9-2 CRB	☐ Challenge 9-2 CRB	☐ *Success for ELL* 9-2
☐ Homework Help Online Keyword: MR7 9-2	☐ Homework Help Online Keyword: MR7 9-2	☐ Homework Help Online Keyword: MR7 9-2	☐ Homework Help Online Keyword: MR7 9-2
☐ *Lesson Tutorial Video* 9-2	☐ *Lesson Tutorial Video* 9-2	☐ *Lesson Tutorial Video* 9-2	☐ *Lesson Tutorial Video* 9-2
☐ Reading Strategies 9-2 CRB	☐ Problem Solving 9-2 CRB	☐ Problem Solving 9-2 CRB	☐ Reading Strategies 9-2 CRB
☐ *Questioning Strategies* pp. 150–151			☐ Lesson Vocabulary SE p. 492
☐ *IDEA Works!* 9-2			☐ *Multilingual Glossary*

ASSESSMENT
☐ Lesson Quiz, TE p. 495 and DT 9-2   ☐ State-Specific Test Prep Online Keyword: MR7 TestPrep

Holt Mathematics

Teacher's Name _____ Class _____ Date _____

Lesson Plan 9-3
Converting Customary Units pp. 496–499   Day _____

Objective Students convert customary units of measure.

> **NCTM Standards:** Understand measurable attributes of objects and the units, systems, and processes of measurement.

Pacing
- ☐ 45-minute Classes: 1 day   ☐ 90-minute Classes: 1/2 day   ☐ Other_____

WARM UP
- ☐ Warm Up TE p. 496 and Daily Transparency 9-3
- ☐ Problem of the Day TE p. 496 and Daily Transparency 9-3
- ☐ Countdown to Testing Transparency Week 20

TEACH
- ☐ Lesson Presentation CD-ROM 9-3
- ☐ Alternate Opener, Explorations Transparency 9-3, TE p. 496, and Exploration 9-3
- ☐ Reaching All Learners TE p. 497
- ☐ Teaching Transparency 9-3
- ☐ *Hands-On Lab Activities* 9-3
- ☐ *Know-It Notebook* 9-3

PRACTICE AND APPLY
- ☐ Example 1: Average: 1–6, 51–59 Advanced: 14–18, 36, 51–59
- ☐ Example 2: Average: 1–12, 40–41, 51–59 Advanced: 14–25, 42–43, 51–59
- ☐ Example 3: Average: 1–28, 37–59 Advanced: 14–59

REACHING ALL LEARNERS – Differentiated Instruction for students with

Developing Knowledge	On-level Knowledge	Advanced Knowledge	English Language Development
☐ Concrete Manipulatives TE p. 497	☐ Concrete Manipulatives TE p. 497	☐ Concrete Manipulatives TE p. 497	☐ Concrete Manipulatives TE p. 497
☐ Practice A 9-3 CRB	☐ Practice B 9-3 CRB	☐ Practice C 9-3 CRB	☐ Practice A, B, or C 9-3 CRB
☐ Reteach 9-3 CRB	☐ Puzzles, Twisters & Teasers 9-3 CRB	☐ Challenge 9-3 CRB	☐ *Success for ELL* 9-3
☐ Homework Help Online Keyword: MR7 9-3	☐ Homework Help Online Keyword: MR7 9-3	☐ Homework Help Online Keyword: MR7 9-3	☐ Homework Help Online Keyword: MR7 9-3
☐ *Lesson Tutorial Video* 9-3	☐ *Lesson Tutorial Video* 9-3	☐ *Lesson Tutorial Video* 9-3	☐ *Lesson Tutorial Video* 9-3
☐ Reading Strategies 9-3 CRB	☐ Problem Solving 9-3 CRB	☐ Problem Solving 9-3 CRB	☐ Reading Strategies 9-3 CRB
☐ *Questioning Strategies* pp. 152–153			
☐ *IDEA Works!* 9-3			☐ *Multilingual Glossary*

ASSESSMENT
- ☐ Lesson Quiz, TE p. 499 and DT 9-3   ☐ State-Specific Test Prep Online Keyword: MR7 TestPrep

Teacher's Name _____ Class _____ Date _____

Lesson Plan 9-4
Converting Metric Units pp. 500–503 Day _____

Objective Students convert metric units of measure.

> **NCTM Standards:** Understand measurable attributes of objects and the units, systems, and processes of measurement.

Pacing
☐ 45-minute Classes: 1 day   ☐ 90-minute Classes: 1/2 day   ☐ Other _____

WARM UP
☐ Warm Up TE p. 500 and Daily Transparency 9-4
☐ Problem of the Day TE p. 500 and Daily Transparency 9-4
☐ Countdown to Testing Transparency Week 20

TEACH
☐ Lesson Presentation CD-ROM 9-4
☐ Alternate Opener, Explorations Transparency 9-4, TE p. 500, and Exploration 9-4
☐ Reaching All Learners TE p. 501
☐ Teaching Transparency 9-4
☐ *Hands-On Lab Activities* 9-4
☐ *Know-It Notebook* 9-4

PRACTICE AND APPLY
☐ Example 1: Average: 1, 40–48 Advanced: 12, 40–48
☐ Example 2: Average: 1–5, 40–48 Advanced: 12–15, 23, 40–48
☐ Example 3: Average: 1–27, 36–48 Advanced: 12–48

REACHING ALL LEARNERS – Differentiated Instruction for students with

Developing Knowledge	On-level Knowledge	Advanced Knowledge	English Language Development
☐ Graphic Organizers TE p. 501	☐ Graphic Organizers TE p. 501	☐ Graphic Organizers TE p. 501	☐ Graphic Organizers TE p. 501
☐ Practice A 9-4 CRB	☐ Practice B 9-4 CRB	☐ Practice C 9-4 CRB	☐ Practice A, B, or C 9-4 CRB
☐ Reteach 9-4 CRB	☐ Puzzles, Twisters & Teasers 9-4 CRB	☐ Challenge 9-4 CRB	☐ *Success for ELL* 9-4
☐ Homework Help Online Keyword: MR7 9-4	☐ Homework Help Online Keyword: MR7 9-4	☐ Homework Help Online Keyword: MR7 9-4	☐ Homework Help Online Keyword: MR7 9-4
☐ *Lesson Tutorial Video* 9-4	☐ *Lesson Tutorial Video* 9-4	☐ *Lesson Tutorial Video* 9-4	☐ *Lesson Tutorial Video* 9-4
☐ Reading Strategies 9-4 CRB	☐ Problem Solving 9-4 CRB	☐ Problem Solving 9-4 CRB	☐ Reading Strategies 9-4 CRB
☐ *Questioning Strategies* pp. 154–155			
☐ *IDEA Works!* 9-4			☐ *Multilingual Glossary*

ASSESSMENT
☐ Lesson Quiz, TE p. 503 and DT 9-4   ☐ State-Specific Test Prep Online Keyword: MR7 TestPrep

Holt Mathematics

Teacher's Name _____ Class _____ Date _____

Lesson Plan 9-5
Time and Temperature pp. 504–507 Day _____

Objective Students find measures of time and temperature.

NCTM Standards: Apply appropriate techniques, tools, and formulas to determine measurements.

Pacing
☐ 45-minute Classes: 1 day ☐ 90-minute Classes: 1/2 day ☐ Other _____

WARM UP
☐ Warm Up TE p. 504 and Daily Transparency 9-5
☐ Problem of the Day TE p. 504 and Daily Transparency 9-5
☐ Countdown to Testing Transparency Week 20

TEACH
☐ Lesson Presentation CD-ROM 9-5
☐ Alternate Opener, Explorations Transparency 9-5, TE p. 504, and Exploration 9-5
☐ Reaching All Learners TE p. 505
☐ Teaching Transparency 9-5
☐ *Know-It Notebook* 9-5

PRACTICE AND APPLY
☐ Example 1: Average: 1–6, 23–24, 35–44 Advanced: 12–17, 24–25, 35–44
☐ Example 2: Average: 1–8, 23–28, 35–44 Advanced: 12–19, 23–27, 31, 35–44
☐ Example 3: Average: 1–11, 18–44 Advanced: 12–44

REACHING ALL LEARNERS – Differentiated Instruction for students with

Developing Knowledge	On-level Knowledge	Advanced Knowledge	English Language Development
☐ Curriculum Integration TE p. 505	☐ Curriculum Integration TE p. 505	☐ Curriculum Integration TE p. 505	☐ Curriculum Integration TE p. 505
☐ Practice A 9-5 CRB	☐ Practice B 9-5 CRB	☐ Practice C 9-5 CRB	☐ Practice A, B, or C 9-5 CRB
☐ Reteach 9-5 CRB	☐ Puzzles, Twisters & Teasers 9-5 CRB	☐ Challenge 9-5 CRB	☐ *Success for ELL* 9-5
☐ Homework Help Online Keyword: MR7 9-5	☐ Homework Help Online Keyword: MR7 9-5	☐ Homework Help Online Keyword: MR7 9-5	☐ Homework Help Online Keyword: MR7 9-5
☐ *Lesson Tutorial Video* 9-5	☐ *Lesson Tutorial Video* 9-5	☐ *Lesson Tutorial Video* 9-5	☐ *Lesson Tutorial Video* 9-5
☐ Reading Strategies 9-5 CRB	☐ Problem Solving 9-5 CRB	☐ Problem Solving 9-5 CRB	☐ Reading Strategies 9-5 CRB
☐ *Questioning Strategies* pp. 156–157			
☐ *IDEA Works!* 9-5			☐ *Multilingual Glossary*

ASSESSMENT
☐ Lesson Quiz, TE p. 507 and DT 9-5   ☐ State-Specific Test Prep Online Keyword: MR7 TestPrep

Holt Mathematics

Teacher's Name _____ Class _____ Date _____

Lesson Plan 9-6
Finding Angle Measures in Polygons pp. 510–513 Day _____

Objective Students find angle measures in polygons.

> **NCTM Standards:** Apply appropriate techniques, tools, and formulas to determine measurements.

Pacing
- [ ] 45-minute Classes: 1 day
- [ ] 90-minute Classes: 1/2 day
- [ ] Other_____

WARM UP
- [ ] Warm Up TE p. 510 and Daily Transparency 9-6
- [ ] Problem of the Day TE p. 510 and Daily Transparency 9-6
- [ ] Countdown to Testing Transparency Week 21

TEACH
- [ ] Lesson Presentation CD-ROM 9-6
- [ ] Alternate Opener, Explorations Transparency 9-6, TE p. 510, and Exploration 9-6
- [ ] Reaching All Learners TE p. 511
- [ ] Teaching Transparency 9-6
- [ ] *Know-It Notebook* 9-6

PRACTICE AND APPLY
- [ ] Example 1: Average: 1–3, 15, 22–28 Advanced: 8–10, 16, 22–28
- [ ] Example 2: Average: 1–6, 15–16, 22–28 Advanced: 8–13, 15–16, 22–28
- [ ] Example 3: Average: 1–12, 17–28 Advanced: 8–28

REACHING ALL LEARNERS – Differentiated Instruction for students with

Developing Knowledge	On-level Knowledge	Advanced Knowledge	English Language Development
☐ Cooperative Learning TE p. 511	☐ Cooperative Learning TE p. 511	☐ Cooperative Learning TE p. 511	☐ Cooperative Learning TE p. 511
☐ Practice A 9-6 CRB	☐ Practice B 9-6 CRB	☐ Practice C 9-6 CRB	☐ Practice A, B, or C 9-6 CRB
☐ Reteach 9-6 CRB	☐ Puzzles, Twisters & Teasers 9-6 CRB	☐ Challenge 9-6 CRB	☐ *Success for ELL* 9-6
☐ Homework Help Online Keyword: MR7 9-6	☐ Homework Help Online Keyword: MR7 9-6	☐ Homework Help Online Keyword: MR7 9-6	☐ Homework Help Online Keyword: MR7 9-6
☐ *Lesson Tutorial Video* 9-6	☐ *Lesson Tutorial Video* 9-6	☐ *Lesson Tutorial Video* 9-6	☐ *Lesson Tutorial Video* 9-6
☐ Reading Strategies 9-6 CRB	☐ Problem Solving 9-6 CRB	☐ Problem Solving 9-6 CRB	☐ Reading Strategies 9-6 CRB
☐ *Questioning Strategies* pp. 158–159			
☐ *IDEA Works!* 9-6			☐ *Multilingual Glossary*

ASSESSMENT
- [ ] Lesson Quiz, TE p. 513 and DT 9-6
- [ ] State-Specific Test Prep Online Keyword: MR7 TestPrep

Teacher's Name _____ Class _____ Date _____

Lesson Plan 9-7
Perimeter pp. 514–517   Day _____

Objective Students find the perimeter and missing side lengths of a polygon.

> **NCTM Standards:** Analyze characteristics and properties of two- and three-dimensional geometric shapes and develop mathematical arguments about geometric relationships; Understand measurable attributes of objects and the units, systems, and processes of measurement.

Pacing
- ☐ 45-minute Classes: 1 day   ☐ 90-minute Classes: 1/2 day   ☐ Other_____

WARM UP
- ☐ Warm Up TE p. 514 and Daily Transparency 9-7
- ☐ Problem of the Day TE p. 514 and Daily Transparency 9-7
- ☐ Countdown to Testing Transparency Week 21

TEACH
- ☐ Lesson Presentation CD-ROM 9-7
- ☐ Alternate Opener, Explorations Transparency 9-7, TE p. 514, and Exploration 9-7
- ☐ Reaching All Learners TE p. 515
- ☐ Teaching Transparency 9-7
- ☐ *Hands-On Lab Activities* 9-7
- ☐ *Technology Lab Activities* 9-7
- ☐ *Know-It Notebook* 9-7

PRACTICE AND APPLY
- ☐ Example 1: Average: 1–2, 16, 23–32 Advanced: 6–7, 19, 23–32
- ☐ Example 2: Average: 1–4, 16–17, 23–32 Advanced: 6–10, 18–19, 23–32
- ☐ Example 3: Average: 1–14, 19–32 Advanced: 6–32

REACHING ALL LEARNERS – Differentiated Instruction for students with

Developing Knowledge	On-level Knowledge	Advanced Knowledge	English Language Development
☐ Modeling TE p. 515	☐ Modeling TE p. 515	☐ Modeling TE p. 515	☐ Modeling TE p. 515
☐ Practice A 9-7 CRB	☐ Practice B 9-7 CRB	☐ Practice C 9-7 CRB	☐ Practice A, B, or C 9-7 CRB
☐ Reteach 9-7 CRB	☐ Puzzles, Twisters & Teasers 9-7 CRB	☐ Challenge 9-7 CRB	☐ *Success for ELL* 9-7
☐ Homework Help Online Keyword: MR7 9-7	☐ Homework Help Online Keyword: MR7 9-7	☐ Homework Help Online Keyword: MR7 9-7	☐ Homework Help Online Keyword: MR7 9-7
☐ *Lesson Tutorial Video* 9-7	☐ *Lesson Tutorial Video* 9-7	☐ *Lesson Tutorial Video* 9-7	☐ *Lesson Tutorial Video* 9-7
☐ Reading Strategies 9-7 CRB	☐ Problem Solving 9-7 CRB	☐ Problem Solving 9-7 CRB	☐ Reading Strategies 9-7 CRB
☐ *Questioning Strategies* pp. 160–161			☐ Lesson Vocabulary SE p. 514
☐ *IDEA Works!* 9-7			☐ *Multilingual Glossary*

ASSESSMENT
- ☐ Lesson Quiz, TE p. 517 and DT 9-7   ☐ State-Specific Test Prep Online Keyword: MR7 TestPrep

Teacher's Name _____ Class _____ Date _____

Lesson Plan 9-8
Circles and Circumference pp. 520–523   Day _____

Objective Students identify the parts of a circle and find the circumference of a circle.

> **NCTM Standards:** Analyze characteristics and properties of two- and three-dimensional geometric shapes and develop mathematical arguments about geometric relationships; Use visualization, spatial reasoning, and geometric modeling to solve problems; Apply appropriate techniques, tools, and formulas to determine measurements.

Pacing
- ☐ 45-minute Classes: 1 day   ☐ 90-minute Classes: 1/2 day   ☐ Other_____

WARM UP
- ☐ Warm Up TE p. 520 and Daily Transparency 9-8
- ☐ Problem of the Day TE p. 520 and Daily Transparency 9-8
- ☐ Countdown to Testing Transparency Week 21

TEACH
- ☐ Lesson Presentation CD-ROM 9-8
- ☐ Alternate Opener, Explorations Transparency 9-8, TE p. 520, and Exploration 9-8
- ☐ Reaching All Learners TE p. 521
- ☐ Teaching Transparency 9-8
- ☐ *Hands-On Lab Activities* 9-8
- ☐ *Know-It Notebook* 9-8

PRACTICE AND APPLY
- ☐ Example 1: Average: 1, 22–32   Advanced: 6, 22–32
- ☐ Example 2: Average: 1–3, 22–32   Advanced: 6–8, 22–32
- ☐ Example 3: Average: 1–12, 17–32   Advanced: 6–32

REACHING ALL LEARNERS – Differentiated Instruction for students with

Developing Knowledge	On-level Knowledge	Advanced Knowledge	English Language Development
☐ Inclusion TE p. 521	☐ Critical Thinking TE p. 521	☐ Critical Thinking TE p. 521	☐ Critical Thinking TE p. 521
☐ Practice A 9-8 CRB	☐ Practice B 9-8 CRB	☐ Practice C 9-8 CRB	☐ Practice A, B, or C 9-8 CRB
☐ Reteach 9-8 CRB	☐ Puzzles, Twisters & Teasers 9-8 CRB	☐ Challenge 9-8 CRB	☐ *Success for ELL* 9-8
☐ Homework Help Online Keyword: MR7 9-8	☐ Homework Help Online Keyword: MR7 9-8	☐ Homework Help Online Keyword: MR7 9-8	☐ Homework Help Online Keyword: MR7 9-8
☐ *Lesson Tutorial Video* 9-8	☐ *Lesson Tutorial Video* 9-8	☐ *Lesson Tutorial Video* 9-8	☐ *Lesson Tutorial Video* 9-8
☐ Reading Strategies 9-8 CRB	☐ Problem Solving 9-8 CRB	☐ Problem Solving 9-8 CRB	☐ Reading Strategies 9-8 CRB
☐ *Questioning Strategies* pp. 162–163			☐ Lesson Vocabulary SE p. 520
☐ *IDEA Works!* 9-8			☐ *Multilingual Glossary*

ASSESSMENT
- ☐ Lesson Quiz, TE p. 523 and DT 9-8   ☐ State-Specific Test Prep Online   Keyword: MR7 TestPrep

Copyright © Holt, Rinehart and Winston.
All rights reserved.

Holt Mathematics

Teacher's Name _____ Class _____ Date _____

Lesson Plan 10-1
Estimating and Finding Area pp. 542–545   Day _____

Objective Students estimate the area of irregular figures and find the area of rectangles and parallelograms.

> **NCTM Standards:** Understand measurable attributes of objects and the units, systems, and processes of measurement; Apply appropriate techniques, tools, and formulas to determine measurements.

Pacing
☐ 45-minute Classes: 1 day   ☐ 90-minute Classes: 1/2 day   ☐ Other _____

WARM UP
☐ Warm Up TE p. 542 and Daily Transparency 10-1
☐ Problem of the Day TE p. 542 and Daily Transparency 10-1
☐ Countdown to Testing Transparency Week 22

TEACH
☐ Lesson Presentation CD-ROM 10-1
☐ Alternate Opener, Explorations Transparency 10-1, TE p. 542, and Exploration 10-1
☐ Reaching All Learners TE p. 543
☐ Teaching Transparency 10-1
☐ *Hands-On Lab Activities* 10-1
☐ *Know-It Notebook* 10-1

PRACTICE AND APPLY
☐ Example 1: Average: 1–3, 25–31 Advanced: 11–13, 25–31
☐ Example 2: Average: 1–6, 25–31 Advanced: 11–16, 25–31
☐ Example 3: Average: 1–9, 25–31 Advanced: 11–19, 25–31
☐ Example 4: Average: 1–10, 20–31 Advanced: 11–31

REACHING ALL LEARNERS – Differentiated Instruction for students with

Developing Knowledge	On-level Knowledge	Advanced Knowledge	English Language Development
☐ Curriculum Integration TE p. 543	☐ Curriculum Integration TE p. 543	☐ Curriculum Integration TE p. 543	☐ Curriculum Integration TE p. 543
☐ Practice A 10-1 CRB	☐ Practice B 10-1 CRB	☐ Practice C 10-1 CRB	☐ Practice A, B, or C 10-1 CRB
☐ Reteach 10-1 CRB	☐ Puzzles, Twisters & Teasers 10-1 CRB	☐ Challenge 10-1 CRB	☐ *Success for ELL* 10-1
☐ Homework Help Online Keyword: MR7 10-1	☐ Homework Help Online Keyword: MR7 10-1	☐ Homework Help Online Keyword: MR7 10-1	☐ Homework Help Online Keyword: MR7 10-1
☐ *Lesson Tutorial Video* 10-1	☐ *Lesson Tutorial Video* 10-1	☐ *Lesson Tutorial Video* 10-1	☐ *Lesson Tutorial Video* 10-1
☐ Reading Strategies 10-1 CRB	☐ Problem Solving 10-1 CRB	☐ Problem Solving 10-1 CRB	☐ Reading Strategies 10-1 CRB
☐ *Questioning Strategies* pp. 164–165			☐ Lesson Vocabulary SE p. 542
☐ *IDEA Works!* 10-1			☐ *Multilingual Glossary*

ASSESSMENT
☐ Lesson Quiz, TE p. 545 and DT 10-1   ☐ State-Specific Test Prep Online Keyword: MR7 TestPrep

Teacher's Name _____ Class _____ Date _____

Lesson Plan 10-2
Area of Triangles and Trapezoids pp. 546–549   Day _____

Objective Students find the area of triangles and trapezoids.

> **NCTM Standards:** Understand measurable attributes of objects and the units, systems, and processes of measurement.

Pacing
☐ 45-minute Classes: 1 day   ☐ 90-minute Classes: 1/2 day   ☐ Other _____

WARM UP
☐ Warm Up TE p. 546 and Daily Transparency 10-2
☐ Problem of the Day TE p. 546 and Daily Transparency 10-2
☐ Countdown to Testing Transparency Week 22

TEACH
☐ Lesson Presentation CD-ROM 10-2
☐ Alternate Opener, Explorations Transparency 10-2, TE p. 546, and Exploration 10-2
☐ Reaching All Learners TE p. 547
☐ Teaching Transparency 10-2
☐ *Know-It Notebook* 10-2

PRACTICE AND APPLY
☐ Example 1: Average: 1–3, 17, 28–34 Advanced: 8–10, 18, 28–34
☐ Example 2: Average: 1–4, 17–18, 20, 28–34 Advanced: 8–12, 17–18, 23, 28–34
☐ Example 3: Average: 1–4, 12–24, 28–34 Advanced: 8–21, 25–34

REACHING ALL LEARNERS – Differentiated Instruction for students with

Developing Knowledge	On-level Knowledge	Advanced Knowledge	English Language Development
☐ Concrete Manipulatives TE p. 547	☐ Concrete Manipulatives TE p. 547	☐ Concrete Manipulatives TE p. 547	☐ Concrete Manipulatives TE p. 547
☐ Practice A 10-2 CRB	☐ Practice B 10-2 CRB	☐ Practice C 10-2 CRB	☐ Practice A, B, or C 10-2 CRB
☐ Reteach 10-2 CRB	☐ Puzzles, Twisters & Teasers 10-2 CRB	☐ Challenge 10-2 CRB	☐ *Success for ELL* 10-2
☐ Homework Help Online Keyword: MR7 10-2	☐ Homework Help Online Keyword: MR7 10-2	☐ Homework Help Online Keyword: MR7 10-2	☐ Homework Help Online Keyword: MR7 10-2
☐ *Lesson Tutorial Video* 10-2	☐ *Lesson Tutorial Video* 10-2	☐ *Lesson Tutorial Video* 10-2	☐ *Lesson Tutorial Video* 10-2
☐ Reading Strategies 10-2 CRB	☐ Problem Solving 10-2 CRB	☐ Problem Solving 10-2 CRB	☐ Reading Strategies 10-2 CRB
☐ *Questioning Strategies* pp. 166–167			
☐ *IDEA Works!* 10-2			☐ *Multilingual Glossary*

ASSESSMENT
☐ Lesson Quiz, TE p. 549 and DT 10-2   ☐ State-Specific Test Prep Online Keyword: MR7 TestPrep

Teacher's Name _____ Class _____ Date _____

Lesson Plan 10-3
Area of Composite Figures pp. 551–553   Day _____

Objective Students break a polygon into simpler parts to find its area.

> **NCTM Standards:** Analyze characteristics and properties of two- and three-dimensional geometric shapes and develop mathematical arguments about geometric relationships; Apply appropriate techniques, tools, and formulas to determine measurements.

Pacing
☐ 45-minute Classes: 1 day   ☐ 90-minute Classes: 1/2 day   ☐ Other_____

WARM UP
☐ Warm Up TE p. 551 and Daily Transparency 10-3
☐ Problem of the Day TE p. 551 and Daily Transparency 10-3
☐ Countdown to Testing Transparency Week 22

TEACH
☐ Lesson Presentation CD-ROM 10-3
☐ Alternate Opener, Explorations Transparency 10-3, TE p. 551, and Exploration 10-3
☐ Reaching All Learners TE p. 552
☐ *Know-It Notebook* 10-3

PRACTICE AND APPLY
☐ Example 1: Average: 1–2, 10–17 Advanced: 4–5, 10–17
☐ Example 2: Average: 1–3, 7–8, 10–17 Advanced: 4–6, 8–17

REACHING ALL LEARNERS – Differentiated Instruction for students with

Developing Knowledge	On-level Knowledge	Advanced Knowledge	English Language Development
☐ Cooperative Learning TE p. 552	☐ Cooperative Learning TE p. 552	☐ Cooperative Learning TE p. 552	☐ Cooperative Learning TE p. 552
☐ Practice A 10-3 CRB	☐ Practice B 10-3 CRB	☐ Practice C 10-3 CRB	☐ Practice A, B, or C 10-3 CRB
☐ Reteach 10-3 CRB	☐ Puzzles, Twisters & Teasers 10-3 CRB	☐ Challenge 10-3 CRB	☐ *Success for ELL* 10-3
☐ Homework Help Online Keyword: MR7 10-3	☐ Homework Help Online Keyword: MR7 10-3	☐ Homework Help Online Keyword: MR7 10-3	☐ Homework Help Online Keyword: MR7 10-3
☐ *Lesson Tutorial Video* 10-3	☐ *Lesson Tutorial Video* 10-3	☐ *Lesson Tutorial Video* 10-3	☐ *Lesson Tutorial Video* 10-3
☐ Reading Strategies 10-3 CRB	☐ Problem Solving 10-3 CRB	☐ Problem Solving 10-3 CRB	☐ Reading Strategies 10-3 CRB
☐ *Questioning Strategies* pp. 168–169	☐ Visual TE p. 551	☐ Visual TE p. 551	
☐ *IDEA Works!* 10-3			☐ *Multilingual Glossary*

ASSESSMENT
☐ Lesson Quiz, TE p. 553 and DT 10-3   ☐ State-Specific Test Prep Online Keyword: MR7 TestPrep

Teacher's Name _____ Class _____ Date _____

Lesson Plan 10-4
Comparing Perimeter and Area pp. 554–556   Day _____

Objective Students make a model to explore how area and perimeter are affected by changes in the dimensions of a figure.

> **NCTM Standards:** Analyze characteristics and properties of two- and three-dimensional geometric shapes and develop mathematical arguments about geometric relationships; Apply appropriate techniques, tools, and formulas to determine measurements.

Pacing
☐ 45-minute Classes: 1 day   ☐ 90-minute Classes: 1/2 day   ☐ Other_____

WARM UP
☐ Warm Up TE p. 554 and Daily Transparency 10-4
☐ Problem of the Day TE p. 554 and Daily Transparency 10-4
☐ Countdown to Testing Transparency Week 22

TEACH
☐ Lesson Presentation CD-ROM 10-4
☐ Alternate Opener, Explorations Transparency 10-4, TE p. 554, and Exploration 10-4
☐ Reaching All Learners TE p. 555
☐ *Hands-On Lab Activities* 10-4
☐ *Know-It Notebook* 10-4

PRACTICE AND APPLY
☐ Example 1: Average: 1, 9–17 Advanced: 3, 9–17
☐ Example 2: Average: 1–6, 9–17 Advanced: 3–17

REACHING ALL LEARNERS – Differentiated Instruction for students with

Developing Knowledge	On-level Knowledge	Advanced Knowledge	English Language Development
☐ Kinesthetic Experience TE p. 555	☐ Kinesthetic Experience TE p. 555	☐ Kinesthetic Experience TE p. 555	☐ Kinesthetic Experience TE p. 555
☐ Practice A 10-4 CRB	☐ Practice B 10-4 CRB	☐ Practice C 10-4 CRB	☐ Practice A, B, or C 10-4 CRB
☐ Reteach 10-4 CRB	☐ Puzzles, Twisters & Teasers 10-4 CRB	☐ Challenge 10-4 CRB	☐ *Success for ELL* 10-4
☐ Homework Help Online Keyword: MR7 10-4	☐ Homework Help Online Keyword: MR7 10-4	☐ Homework Help Online Keyword: MR7 10-4	☐ Homework Help Online Keyword: MR7 10-4
☐ *Lesson Tutorial Video* 10-4	☐ *Lesson Tutorial Video* 10-4	☐ *Lesson Tutorial Video* 10-4	☐ *Lesson Tutorial Video* 10-4
☐ Reading Strategies 10-4 CRB	☐ Problem Solving 10-4 CRB	☐ Problem Solving 10-4 CRB	☐ Reading Strategies 10-4 CRB
☐ *Questioning Strategies* pp. 170–171			
☐ *IDEA Works!* 10-4			☐ *Multilingual Glossary*

ASSESSMENT
☐ Lesson Quiz, TE p. 556 and DT 10-4   ☐ State-Specific Test Prep Online Keyword: MR7 TestPrep

Copyright © Holt, Rinehart and Winston.
All rights reserved.

Holt Mathematics

Teacher's Name _____ Class _____ Date _____

Lesson Plan 10-5
Area of Circles pp. 558–561   Day _____

Objective Students find the area of a circle.

> **NCTM Standards:** Understand measurable attributes of objects and the units, systems, and processes of measurement.

Pacing
☐ 45-minute Classes: 1 day   ☐ 90-minute Classes: 1/2 day   ☐ Other _____

WARM UP
☐ Warm Up TE p. 558 and Daily Transparency 10-5
☐ Problem of the Day TE p. 558 and Daily Transparency 10-5
☐ Countdown to Testing Transparency Week 23

TEACH
☐ Lesson Presentation CD-ROM 10-5
☐ Alternate Opener, Explorations Transparency 10-5, TE p. 558, and Exploration 10-5
☐ Reaching All Learners TE p. 559
☐ Teaching Transparency 10-5
☐ *Know-It Notebook* 10-5

PRACTICE AND APPLY
☐ Example 1: Average: 1–3, 25–34 Advanced: 8–10, 25–34
☐ Example 2: Average: 1–6, 15–16, 25–34 Advanced: 8–13, 16–17, 25–34
☐ Example 3: Average: 1–12, 18–21, 25–34 Advanced: 8–34

REACHING ALL LEARNERS – Differentiated Instruction for students with

Developing Knowledge	On-level Knowledge	Advanced Knowledge	English Language Development
☐ Multiple Representations TE p. 559	☐ Multiple Representations TE p. 559	☐ Multiple Representations TE p. 559	☐ Multiple Representations TE p. 559
☐ Practice A 10-5 CRB	☐ Practice B 10-5 CRB	☐ Practice C 10-5 CRB	☐ Practice A, B, or C 10-5 CRB
☐ Reteach 10-5 CRB	☐ Puzzles, Twisters & Teasers 10-5 CRB	☐ Challenge 10-5 CRB	☐ *Success for ELL* 10-5
☐ Homework Help Online Keyword: MR7 10-5	☐ Homework Help Online Keyword: MR7 10-5	☐ Homework Help Online Keyword: MR7 10-5	☐ Homework Help Online Keyword: MR7 10-5
☐ *Lesson Tutorial Video* 10-5	☐ *Lesson Tutorial Video* 10-5	☐ *Lesson Tutorial Video* 10-5	☐ *Lesson Tutorial Video* 10-5
☐ Reading Strategies 10-5 CRB	☐ Problem Solving 10-5 CRB	☐ Problem Solving 10-5 CRB	☐ Reading Strategies 10-5 CRB
☐ *Questioning Strategies* pp. 172–173			
☐ *IDEA Works!* 10-5			☐ *Multilingual Glossary*

ASSESSMENT
☐ Lesson Quiz, TE p. 561 and DT 10-5   ☐ State-Specific Test Prep Online Keyword: MR7 TestPrep

Teacher's Name _____ Class _____ Date _____

Lesson Plan 10-6
Three-Dimensional Figures pp. 566–569   Day _____

Objective Students name three-dimensional figures.

> **NCTM Standards:** Analyze characteristics and properties of two- and three-dimensional geometric shapes and develop mathematical arguments about geometric relationships; Use visualization, spatial reasoning, and geometric modeling to solve problems.

Pacing
- [ ] 45-minute Classes: 1 day   - [ ] 90-minute Classes: 1/2 day   - [ ] Other_____

WARM UP
- [ ] Warm Up TE p. 566 and Daily Transparency 10-6
- [ ] Problem of the Day TE p. 566 and Daily Transparency 10-6
- [ ] Countdown to Testing Transparency Week 23

TEACH
- [ ] Lesson Presentation CD-ROM 10-6
- [ ] Alternate Opener, Explorations Transparency 10-6, TE p. 566, and Exploration 10-6
- [ ] Reaching All Learners TE p. 567
- [ ] Teaching Transparency 10-6
- [ ] *Hands-On Lab Activities* 10-6
- [ ] *Know-It Notebook* 10-6

PRACTICE AND APPLY
- [ ] Example 1: Average: 1–3, 18, 29–37 Advanced: 7–9, 19, 29–37
- [ ] Example 2: Average: 1–13, 20–26, 29–37 Advanced: 7–37

REACHING ALL LEARNERS – Differentiated Instruction for students with

Developing Knowledge	On-level Knowledge	Advanced Knowledge	English Language Development
☐ Critical Thinking TE p. 567	☐ Critical Thinking TE p. 567	☐ Critical Thinking TE p. 567	☐ Critical Thinking TE p. 567
☐ Practice A 10-6 CRB	☐ Practice B 10-6 CRB	☐ Practice C 10-6 CRB	☐ Practice A, B, or C 10-6 CRB
☐ Reteach 10-6 CRB	☐ Puzzles, Twisters & Teasers 10-6 CRB	☐ Challenge 10-6 CRB	☐ *Success for ELL* 10-6
☐ Homework Help Online Keyword: MR7 10-6	☐ Homework Help Online Keyword: MR7 10-6	☐ Homework Help Online Keyword: MR7 10-6	☐ Homework Help Online Keyword: MR7 10-6
☐ *Lesson Tutorial Video* 10-6	☐ *Lesson Tutorial Video* 10-6	☐ *Lesson Tutorial Video* 10-6	☐ *Lesson Tutorial Video* 10-6
☐ Reading Strategies 10-6 CRB	☐ Problem Solving 10-6 CRB	☐ Problem Solving 10-6 CRB	☐ Reading Strategies 10-6 CRB
☐ *Questioning Strategies* pp. 174–175			☐ Lesson Vocabulary SE p. 566
☐ *IDEA Works!* 10-6			☐ *Multilingual Glossary*

ASSESSMENT
- [ ] Lesson Quiz, TE p. 569 and DT 10-6   - [ ] State-Specific Test Prep Online Keyword: MR7 TestPrep

Copyright © Holt, Rinehart and Winston.
All rights reserved.

Holt Mathematics

Teacher's Name _____ Class _____ Date _____

Lesson Plan 10-7
Volume of Prisms pp. 572–575 Day _____

Objective Students estimate and find the volumes of rectangular prisms and triangular prisms.

> **NCTM Standards:** Analyze characteristics and properties of two- and three-dimensional geometric shapes and develop mathematical arguments about geometric relationships; Understand measurable attributes of objects and the units, systems, and processes of measurement.

Pacing
☐ 45-minute Classes: 1 day ☐ 90-minute Classes: 1/2 day ☐ Other_____

WARM UP
☐ Warm Up TE p. 572 and Daily Transparency 10-7
☐ Problem of the Day TE p. 572 and Daily Transparency 10-7
☐ Countdown to Testing Transparency Week 24

TEACH
☐ Lesson Presentation CD-ROM 10-7
☐ Alternate Opener, Explorations Transparency 10-7, TE p. 572, and Exploration 10-7
☐ Reaching All Learners TE p. 573
☐ Teaching Transparency 10-7
☐ *Hands-On Lab Activities* 10-7
☐ *Know-It Notebook* 10-7

PRACTICE AND APPLY
☐ Example 1: Average: 1–3, 15–16, 28–33 Advanced: 8–10, 15–16, 28–33
☐ Example 2: Average: 1–6, 15–19, 21, 28–33 Advanced: 8–13, 15–20, 28–33
☐ Example 2: Average: 1–16, 21–25, 28–33 Advanced: 8–33

REACHING ALL LEARNERS – Differentiated Instruction for students with

Developing Knowledge	On-level Knowledge	Advanced Knowledge	English Language Development
☐ Cooperative Learning TE p. 573	☐ Cooperative Learning TE p. 573	☐ Cooperative Learning TE p. 573	☐ Cooperative Learning TE p. 573
☐ Practice A 10-7 CRB	☐ Practice B 10-7 CRB	☐ Practice C 10-7 CRB	☐ Practice A, B, or C 10-7 CRB
☐ Reteach 10-7 CRB	☐ Puzzles, Twisters & Teasers 10-7 CRB	☐ Challenge 10-7 CRB	☐ *Success for ELL* 10-7
☐ Homework Help Online Keyword: MR7 10-7	☐ Homework Help Online Keyword: MR7 10-7	☐ Homework Help Online Keyword: MR7 10-7	☐ Homework Help Online Keyword: MR7 10-7
☐ *Lesson Tutorial Video* 10-7	☐ *Lesson Tutorial Video* 10-7	☐ *Lesson Tutorial Video* 10-7	☐ *Lesson Tutorial Video* 10-7
☐ Reading Strategies 10-7 CRB	☐ Problem Solving 10-7 CRB	☐ Problem Solving 10-7 CRB	☐ Reading Strategies 10-7 CRB
☐ *Questioning Strategies* pp. 176–177			☐ Lesson Vocabulary SE p. 572
☐ *IDEA Works!* 10-7			☐ *Multilingual Glossary*

ASSESSMENT
☐ Lesson Quiz, TE p. 575 and DT 10-7   ☐ State-Specific Test Prep Online Keyword: MR7 TestPrep

Teacher's Name _____ Class _____ Date _____

Lesson Plan 10-8
Volume of Cylinders pp. 576–579 Day _____

Objective Students find volumes of cylinders.

> **NCTM Standards:** Analyze characteristics and properties of two- and three-dimensional geometric shapes and develop mathematical arguments about geometric relationships; Understand measurable attributes of objects and the units, systems, and processes of measurement; Apply appropriate techniques, tools, and formulas to determine measurements.

Pacing
- ☐ 45-minute Classes: 1 day   ☐ 90-minute Classes: 1/2 day   ☐ Other_____

WARM UP
- ☐ Warm Up TE p. 576 and Daily Transparency 10-8
- ☐ Problem of the Day TE p. 576 and Daily Transparency 10-8
- ☐ Countdown to Testing Transparency Week 24

TEACH
- ☐ Lesson Presentation CD-ROM 10-8
- ☐ Alternate Opener, Explorations Transparency 10-8, TE p. 576, and Exploration 10-8
- ☐ Reaching All Learners TE p. 577
- ☐ *Hands-On Lab Activities* 10-8
- ☐ *Technology Lab Activities* 10-8
- ☐ *Know-It Notebook* 10-8

PRACTICE AND APPLY
- ☐ Example 1: Average: 1–3, 14–16, 28–33 Advanced: 6–8, 17–19, 28–33
- ☐ Example 2: Average: 1–4, 14–19, 28–33 Advanced: 6–9, 14–19, 28–33
- ☐ Example 3: Average: 1–11, 17–33 Advanced: 6–33

REACHING ALL LEARNERS – Differentiated Instruction for students with

Developing Knowledge	On-level Knowledge	Advanced Knowledge	English Language Development
☐ Inclusion TE p. 577	☐ Modeling TE p. 577	☐ Modeling TE p. 577	☐ Modeling TE p. 577
☐ Practice A 10-8 CRB	☐ Practice B 10-8 CRB	☐ Practice C 10-8 CRB	☐ Practice A, B, or C 10-8 CRB
☐ Reteach 10-8 CRB	☐ Puzzles, Twisters & Teasers 10-8 CRB	☐ Challenge 10-8 CRB	☐ *Success for ELL* 10-8
☐ Homework Help Online Keyword: MR7 10-8	☐ Homework Help Online Keyword: MR7 10-8	☐ Homework Help Online Keyword: MR7 10-8	☐ Homework Help Online Keyword: MR7 10-8
☐ *Lesson Tutorial Video* 10-8	☐ *Lesson Tutorial Video* 10-8	☐ *Lesson Tutorial Video* 10-8	☐ *Lesson Tutorial Video* 10-8
☐ Reading Strategies 10-8 CRB	☐ Problem Solving 10-8 CRB	☐ Problem Solving 10-8 CRB	☐ Reading Strategies 10-8 CRB
☐ *Questioning Strategies* pp. 178–179			
☐ *IDEA Works!* 10-8			☐ *Multilingual Glossary*

ASSESSMENT
- ☐ Lesson Quiz, TE p. 579 and DT 10-8   ☐ State-Specific Test Prep Online Keyword: MR7 TestPrep

Teacher's Name _____ Class _____ Date _____

Lesson Plan 10-9

Surface Area pp. 582–585   Day _____

Objective Students find the surface areas of prisms, pyramids, and cylinders.

> **NCTM Standards:** Analyze characteristics and properties of two- and three-dimensional geometric shapes and develop mathematical arguments about geometric relationships; Apply appropriate techniques, tools, and formulas to determine measurements.

Pacing
☐ 45-minute Classes: 1 day   ☐ 90-minute Classes: 1/2 day   ☐ Other_____

WARM UP
☐ Warm Up TE p. 582 and Daily Transparency 10-9
☐ Problem of the Day TE p. 582 and Daily Transparency 10-9
☐ Countdown to Testing Transparency Week 24

TEACH
☐ Lesson Presentation CD-ROM 10-9
☐ Alternate Opener, Explorations Transparency 10-9, TE p. 582, and Exploration 10-9
☐ Reaching All Learners TE p. 583
☐ *Hands-On Lab Activities* 10-9
☐ *Know-It Notebook* 10-9

PRACTICE AND APPLY
☐ Example 1: Average: 1–3, 23, 28–36 Advanced: 10–12, 23, 28–36
☐ Example 2: Average: 1–6, 23, 28–36 Advanced: 10–15, 23, 28–36
☐ Example 3: Average: 1–12, 22–36 Advanced: 10–36

REACHING ALL LEARNERS – Differentiated Instruction for students with

Developing Knowledge	On-level Knowledge	Advanced Knowledge	English Language Development
☐ Inclusion TE p. 583	☐ Modeling TE p. 583	☐ Modeling TE p. 583	☐ Modeling TE p. 583
☐ Practice A 10-9 CRB	☐ Practice B 10-9 CRB	☐ Practice C 10-9 CRB	☐ Practice A, B, or C 10-9 CRB
☐ Reteach 10-9 CRB	☐ Puzzles, Twisters & Teasers 10-9 CRB	☐ Challenge 10-9 CRB	☐ *Success for ELL* 10-9
☐ Homework Help Online Keyword: MR7 10-9	☐ Homework Help Online Keyword: MR7 10-9	☐ Homework Help Online Keyword: MR7 10-9	☐ Homework Help Online Keyword: MR7 10-9
☐ *Lesson Tutorial Video* 10-9	☐ *Lesson Tutorial Video* 10-9	☐ *Lesson Tutorial Video* 10-9	☐ *Lesson Tutorial Video* 10-9
☐ Reading Strategies 10-9 CRB	☐ Problem Solving 10-9 CRB	☐ Problem Solving 10-9 CRB	☐ Reading Strategies 10-9 CRB
☐ *Questioning Strategies* pp. 180–181			☐ Lesson Vocabulary SE p. 582
☐ *IDEA Works!* 10-9			☐ *Multilingual Glossary*

ASSESSMENT
☐ Lesson Quiz, TE p. 585 and DT 10-9   ☐ State-Specific Test Prep Online Keyword: MR7 TestPrep

Teacher's Name _____ Class _____ Date _____

Lesson Plan 11-1
Integers in Real-World Situations pp. 602–605   Day _____

Objective Students identify and graph integers and find opposites of an integer.

> **NCTM Standards:** Understand numbers, ways of representing numbers, relationships among numbers, and number systems; Recognize and apply mathematics in contexts outside of mathematics.

Pacing
- ☐ 45-minute Classes: 1 day
- ☐ 90-minute Classes: 1/2 day
- ☐ Other _____

WARM UP
- ☐ Warm Up TE p. 602 and Daily Transparency 11-1
- ☐ Problem of the Day TE p. 602 and Daily Transparency 11-1

TEACH
- ☐ Lesson Presentation CD-ROM 11-1
- ☐ Alternate Opener, Explorations Transparency 11-1, TE p. 602, and Exploration 11-1
- ☐ Reaching All Learners TE p. 603
- ☐ Teaching Transparency 11-1
- ☐ *Know-It Notebook* 11-1

PRACTICE AND APPLY
- ☐ Example 1: Average: 1–2, 29, 35, 43–48   Advanced: 8–11, 29, 43–48
- ☐ Example 2: Average: 1–6, 18–22, 29, 35–36, 43–48   Advanced: 8–16, 21–22, 29, 37–38, 43–48
- ☐ Example 3: Average: 1–29, 35–48   Advanced: 8–48

REACHING ALL LEARNERS – Differentiated Instruction for students with

Developing Knowledge	On-level Knowledge	Advanced Knowledge	English Language Development
☐ Curriculum Integration TE p. 603	☐ Curriculum Integration TE p. 603	☐ Curriculum Integration TE p. 603	☐ Curriculum Integration TE p. 603
☐ Practice A 11-1 CRB	☐ Practice B 11-1 CRB	☐ Practice C 11-1 CRB	☐ Practice A, B, or C 11-1 CRB
☐ Reteach 11-1 CRB	☐ Puzzles, Twisters & Teasers 11-1 CRB	☐ Challenge 11-1 CRB	☐ *Success for ELL* 11-1
☐ Homework Help Online Keyword: MR7 11-1	☐ Homework Help Online Keyword: MR7 11-1	☐ Homework Help Online Keyword: MR7 11-1	☐ Homework Help Online Keyword: MR7 11-1
☐ *Lesson Tutorial Video* 11-1	☐ *Lesson Tutorial Video* 11-1	☐ *Lesson Tutorial Video* 11-1	☐ *Lesson Tutorial Video* 11-1
☐ Reading Strategies 11-1 CRB	☐ Problem Solving 11-1 CRB	☐ Problem Solving 11-1 CRB	☐ Reading Strategies 11-1 CRB
☐ *Questioning Strategies* pp. 182–183	☐ Reading Math TE p. 603	☐ Reading Math TE p. 603	☐ Lesson Vocabulary SE p. 602
☐ *IDEA Works!* 11-1			☐ *Multilingual Glossary*

ASSESSMENT
- ☐ Lesson Quiz, TE p. 605 and DT 11-1   ☐ State-Specific Test Prep Online Keyword: MR7 TestPrep

Teacher's Name _____ Class _____ Date _____

Lesson Plan 11-2
Comparing and Ordering Integers pp. 606–609   Day _____

Objective Students compare and order integers.

> **NCTM Standards:** Understand numbers, ways of representing numbers, relationships among numbers, and number systems.

Pacing
☐ 45-minute Classes: 1 day   ☐ 90-minute Classes: 1/2 day   ☐ Other _____

WARM UP
☐ Warm Up TE p. 606 and Daily Transparency 11-2
☐ Problem of the Day TE p. 606 and Daily Transparency 11-2

TEACH
☐ Lesson Presentation CD-ROM 11-2
☐ Alternate Opener, Explorations Transparency 11-2, TE p. 606, and Exploration 11-2
☐ Reaching All Learners TE p. 607
☐ *Hands-On Lab Activities* 11-2
☐ *Know-It Notebook* 11-2

PRACTICE AND APPLY
☐ Example 1: Average: 1–3, 19–26, 34, 39–49 Advanced: 8–11, 19–26, 36, 39–49
☐ Example 2: Average: 1–6, 19–35, 39–49 Advanced: 8–17, 19–33, 39–49
☐ Example 3: Average: 1–22, 29–49 Advanced: 8–49

REACHING ALL LEARNERS – Differentiated Instruction for students with

Developing Knowledge	On-level Knowledge	Advanced Knowledge	English Language Development
☐ Kinesthetic Experience TE p. 607	☐ Kinesthetic Experience TE p. 607	☐ Kinesthetic Experience TE p. 607	☐ Kinesthetic Experience TE p. 607
☐ Practice A 11-2 CRB	☐ Practice B 11-2 CRB	☐ Practice C 11-2 CRB	☐ Practice A, B, or C 11-2 CRB
☐ Reteach 11-2 CRB	☐ Puzzles, Twisters & Teasers 11-2 CRB	☐ Challenge 11-2 CRB	☐ *Success for ELL* 11-2
☐ Homework Help Online Keyword: MR7 11-2	☐ Homework Help Online Keyword: MR7 11-2	☐ Homework Help Online Keyword: MR7 11-2	☐ Homework Help Online Keyword: MR7 11-2
☐ *Lesson Tutorial Video* 11-2	☐ *Lesson Tutorial Video* 11-2	☐ *Lesson Tutorial Video* 11-2	☐ *Lesson Tutorial Video* 11-2
☐ Reading Strategies 11-2 CRB	☐ Problem Solving 11-2 CRB	☐ Problem Solving 11-2 CRB	☐ Reading Strategies 11-2 CRB
☐ *Questioning Strategies* pp. 184–185	☐ Visual TE p. 607	☐ Visual TE p. 607	
☐ *IDEA Works!* 11-2			☐ *Multilingual Glossary*

ASSESSMENT
☐ Lesson Quiz, TE p. 609 and DT 11-2   ☐ State-Specific Test Prep Online Keyword: MR7 TestPrep

Teacher's Name _____ Class _____ Date _____

Lesson Plan 11-3
The Coordinate Plane pp. 610–613 Day _____

Objective Students locate and graph points on a coordinate plane.

> **NCTM Standards:** Represent and analyze mathematical situations and structures using algebraic symbols; Specify locations and describe spatial relationships using coordinate geometry and other representational systems.

Pacing
☐ 45-minute Classes: 1 day   ☐ 90-minute Classes: 1/2 day   ☐ Other_____

WARM UP
☐ Warm Up TE p. 610 and Daily Transparency 11-3
☐ Problem of the Day TE p. 610 and Daily Transparency 11-3

TEACH
☐ Lesson Presentation CD-ROM 11-3
☐ Alternate Opener, Explorations Transparency 11-3, TE p. 610, and Exploration 11-3
☐ Reaching All Learners TE p. 611
☐ Teaching Transparency 11-3
☐ *Technology Lab Activities* 11-3
☐ *Know-It Notebook* 11-3

PRACTICE AND APPLY
☐ Example 1: Average: 1–3, 10–15, 55–65 Advanced: 1–3, 10–15, 55–65
☐ Example 2: Average: 1–6, 10–21, 28–35, 50, 55–65 Advanced: 1–6, 10–21, 28–35, 50, 55–65
☐ Example 3: Average: 1–51, 55–65 Advanced: 1–48, 52–65

REACHING ALL LEARNERS – Differentiated Instruction for students with

Developing Knowledge	On-level Knowledge	Advanced Knowledge	English Language Development
☐ Inclusion TE p. 611	☐ Multiple Representations TE p. 611	☐ Multiple Representations TE p. 611	☐ Multiple Representations TE p. 611
☐ Practice A 11-3 CRB	☐ Practice B 11-3 CRB	☐ Practice C 11-3 CRB	☐ Practice A, B, or C 11-3 CRB
☐ Reteach 11-3 CRB	☐ Puzzles, Twisters & Teasers 11-3 CRB	☐ Challenge 11-3 CRB	☐ *Success for ELL* 11-3
☐ Homework Help Online Keyword: MR7 11-3	☐ Homework Help Online Keyword: MR7 11-3	☐ Homework Help Online Keyword: MR7 11-3	☐ Homework Help Online Keyword: MR7 11-3
☐ *Lesson Tutorial Video* 11-3	☐ *Lesson Tutorial Video* 11-3	☐ *Lesson Tutorial Video* 11-3	☐ *Lesson Tutorial Video* 11-3
☐ Reading Strategies 11-3 CRB	☐ Problem Solving 11-3 CRB	☐ Problem Solving 11-3 CRB	☐ Reading Strategies 11-3 CRB
☐ *Questioning Strategies* pp. 186–187			☐ Lesson Vocabulary SE p. 610
☐ *IDEA Works!* 11-3			☐ *Multilingual Glossary*

ASSESSMENT
☐ Lesson Quiz, TE p. 613 and DT 11-3   ☐ State-Specific Test Prep Online Keyword: MR7 TestPrep

Teacher's Name _____ Class _____ Date _____

Lesson Plan 11-4
Adding Integers pp. 617–620   Day _____

Objective Students add integers.

> **NCTM Standards:** Compute fluently and make reasonable estimates.

Pacing
☐ 45-minute Classes: 1 day   ☐ 90-minute Classes: 1/2 day   ☐ Other_____

WARM UP
☐ Warm Up TE p. 617 and Daily Transparency 11-4
☐ Problem of the Day TE p. 617 and Daily Transparency 11-4

TEACH
☐ Lesson Presentation CD-ROM 11-4
☐ Alternate Opener, Explorations Transparency 11-4, TE p. 617, and Exploration 11-4
☐ Reaching All Learners TE p. 618
☐ Teaching Transparency 11-4
☐ *Hands-On Lab Activities* 11-4
☐ *Know-It Notebook* 11-4

PRACTICE AND APPLY
☐ Example 1: Average: 1, 35–41, 63–75 Advanced: 15–16, 35–42, 63–75
☐ Example 2: Average: 1–7, 35–50, 63–75 Advanced: 15–24, 35–50, 63–75
☐ Example 3: Average: 1–13, 35–56, 63–75 Advanced: 15–32, 40–56, 63–75
☐ Example 4: Average: 1–42, 51–59, 63–75 Advanced: 15–75

REACHING ALL LEARNERS – Differentiated Instruction for students with

Developing Knowledge	On-level Knowledge	Advanced Knowledge	English Language Development
☐ Concrete Manipulatives TE p. 618	☐ Concrete Manipulatives TE p. 618	☐ Concrete Manipulatives TE p. 618	☐ Concrete Manipulatives TE p. 618
☐ Practice A 11-4 CRB	☐ Practice B 11-4 CRB	☐ Practice C 11-4 CRB	☐ Practice A, B, or C 11-4 CRB
☐ Reteach 11-4 CRB	☐ Puzzles, Twisters & Teasers 11-4 CRB	☐ Challenge 11-4 CRB	☐ *Success for ELL* 11-4
☐ Homework Help Online Keyword: MR7 11-4	☐ Homework Help Online Keyword: MR7 11-4	☐ Homework Help Online Keyword: MR7 11-4	☐ Homework Help Online Keyword: MR7 11-4
☐ *Lesson Tutorial Video* 11-4	☐ *Lesson Tutorial Video* 11-4	☐ *Lesson Tutorial Video* 11-4	☐ *Lesson Tutorial Video* 11-4
☐ Reading Strategies 11-4 CRB	☐ Problem Solving 11-4 CRB	☐ Problem Solving 11-4 CRB	☐ Reading Strategies 11-4 CRB
☐ *Questioning Strategies* pp. 188–189			
☐ *IDEA Works!* 11-4			☐ *Multilingual Glossary*

ASSESSMENT
☐ Lesson Quiz, TE p. 620 and DT 11-4   ☐ State-Specific Test Prep Online Keyword: MR7 TestPrep

Teacher's Name _____ Class _____ Date _____

Lesson Plan 11-5
Subtracting Integers pp. 622–624 Day _____

Objective Students subtract integers.

> **NCTM Standards:** Compute fluently and make reasonable estimates.

Pacing
☐ 45-minute Classes: 1 day   ☐ 90-minute Classes: 1/2 day   ☐ Other_____

WARM UP
☐ Warm Up TE p. 622 and Daily Transparency 11-5
☐ Problem of the Day TE p. 622 and Daily Transparency 11-5

TEACH
☐ Lesson Presentation CD-ROM 11-5
☐ Alternate Opener, Explorations Transparency 11-5, TE p. 622, and Exploration 11-5
☐ Reaching All Learners TE p. 623
☐ Teaching Transparency 11-5
☐ *Hands-On Lab Activities* 11-5
☐ *Know-It Notebook* 11-5

PRACTICE AND APPLY
☐ Example 1: Average: 1, 43–54 Advanced: 10, 43–54
☐ Example 2: Average: 1–5, 23–30, 43–54 Advanced: 10–14, 23–30, 43–54
☐ Example 3: Average: 1–39, 43–54 Advanced: 10–54

REACHING ALL LEARNERS – Differentiated Instruction for students with

Developing Knowledge	On-level Knowledge	Advanced Knowledge	English Language Development
☐ Curriculum Integration TE p. 623	☐ Curriculum Integration TE p. 623	☐ Curriculum Integration TE p. 623	☐ Curriculum Integration TE p. 623
☐ Practice A 11-5 CRB	☐ Practice B 11-5 CRB	☐ Practice C 11-5 CRB	☐ Practice A, B, or C 11-5 CRB
☐ Reteach 11-5 CRB	☐ Puzzles, Twisters & Teasers 11-5 CRB	☐ Challenge 11-5 CRB	☐ *Success for ELL* 11-5
☐ Homework Help Online Keyword: MR7 11-5	☐ Homework Help Online Keyword: MR7 11-5	☐ Homework Help Online Keyword: MR7 11-5	☐ Homework Help Online Keyword: MR7 11-5
☐ *Lesson Tutorial Video* 11-5	☐ *Lesson Tutorial Video* 11-5	☐ *Lesson Tutorial Video* 11-5	☐ *Lesson Tutorial Video* 11-5
☐ Reading Strategies 11-5 CRB	☐ Problem Solving 11-5 CRB	☐ Problem Solving 11-5 CRB	☐ Reading Strategies 11-5 CRB
☐ *Questioning Strategies* pp. 190–191			
☐ *IDEA Works!* 11-5			☐ *Multilingual Glossary*

ASSESSMENT
☐ Lesson Quiz, TE p. 624 and DT 11-5   ☐ State-Specific Test Prep Online Keyword: MR7 TestPrep

Teacher's Name _____ Class _____ Date _____

Lesson Plan 11-6
Multiplying Integers pp. 625–627   Day _____

Objective Students multiply integers.

> **NCTM Standards:** Compute fluently and make reasonable estimates.

Pacing
☐ 45-minute Classes: 1 day ☐ 90-minute Classes: 1/2 day ☐ Other _____

WARM UP
☐ Warm Up TE p. 625 and Daily Transparency 11-6
☐ Problem of the Day TE p. 625 and Daily Transparency 11-6

TEACH
☐ Lesson Presentation CD-ROM 11-6
☐ Alternate Opener, Explorations Transparency 11-6, TE p. 625, and Exploration 11-6
☐ Reaching All Learners TE p. 626
☐ Teaching Transparency 11-6
☐ *Technology Lab Activities* 11-6
☐ *Know-It Notebook* 11-6

PRACTICE AND APPLY
☐ Example 1: Average: 1–8, 44–51 Advanced: 17–24, 44–51
☐ Example 2: Average: 1–24, 44–51 Advanced: 17–38, 41–51

REACHING ALL LEARNERS – Differentiated Instruction for students with

Developing Knowledge	On-level Knowledge	Advanced Knowledge	English Language Development
☐ Cooperative Learning TE p. 626	☐ Cooperative Learning TE p. 626	☐ Cooperative Learning TE p. 626	☐ Cooperative Learning TE p. 626
☐ Practice A 11-6 CRB	☐ Practice B 11-6 CRB	☐ Practice C 11-6 CRB	☐ Practice A, B, or C 11-6 CRB
☐ Reteach 11-6 CRB	☐ Puzzles, Twisters & Teasers 11-6 CRB	☐ Challenge 11-6 CRB	☐ *Success for ELL* 11-6
☐ Homework Help Online Keyword: MR7 11-6	☐ Homework Help Online Keyword: MR7 11-6	☐ Homework Help Online Keyword: MR7 11-6	☐ Homework Help Online Keyword: MR7 11-6
☐ *Lesson Tutorial Video* 11-6	☐ *Lesson Tutorial Video* 11-6	☐ *Lesson Tutorial Video* 11-6	☐ *Lesson Tutorial Video* 11-6
☐ Reading Strategies 11-6 CRB	☐ Problem Solving 11-6 CRB	☐ Problem Solving 11-6 CRB	☐ Reading Strategies 11-6 CRB
☐ *Questioning Strategies* pp. 192–193			
☐ *IDEA Works!* 11-6			☐ *Multilingual Glossary*

ASSESSMENT
☐ Lesson Quiz, TE p. 627 and DT 11-6   ☐ State-Specific Test Prep Online Keyword: MR7 TestPrep

Teacher's Name _____ Class _____ Date _____

Lesson Plan 11-7
Dividing Integers pp. 628–631   Day _____

Objective Students divide integers.

> **NCTM Standards:** Compute fluently and make reasonable estimates.

Pacing
- ☐ 45-minute Classes: 1 day   ☐ 90-minute Classes: 1/2 day   ☐ Other _____

WARM UP
- ☐ Warm Up TE p. 628 and Daily Transparency 11-7
- ☐ Problem of the Day TE p. 628 and Daily Transparency 11-7

TEACH
- ☐ Lesson Presentation CD-ROM 11-7
- ☐ Alternate Opener, Explorations Transparency 11-7, TE p. 628, and Exploration 11-7
- ☐ Reaching All Learners TE p. 629
- ☐ Teaching Transparency 11-7
- ☐ *Technology Lab Activities* 11-7
- ☐ *Know-It Notebook* 11-7

PRACTICE AND APPLY
- ☐ Example 1: Average: 1–4, 9–12, 25–28, 48–56 Advanced: 9–16, 29–32, 48–56
- ☐ Example 2: Average: 1–20, 25–28, 33–44, 48–56 Advanced: 9–41, 45–56

REACHING ALL LEARNERS – Differentiated Instruction for students with

Developing Knowledge	On-level Knowledge	Advanced Knowledge	English Language Development
☐ Critical Thinking TE p. 629	☐ Critical Thinking TE p. 629	☐ Critical Thinking TE p. 629	☐ Critical Thinking TE p. 629
☐ Practice A 11-7 CRB	☐ Practice B 11-7 CRB	☐ Practice C 11-7 CRB	☐ Practice A, B, or C 11-7 CRB
☐ Reteach 11-7 CRB	☐ Puzzles, Twisters & Teasers 11-7 CRB	☐ Challenge 11-7 CRB	☐ *Success for ELL* 11-7
☐ Homework Help Online Keyword: MR7 11-7	☐ Homework Help Online Keyword: MR7 11-7	☐ Homework Help Online Keyword: MR7 11-7	☐ Homework Help Online Keyword: MR7 11-7
☐ *Lesson Tutorial Video* 11-7	☐ *Lesson Tutorial Video* 11-7	☐ *Lesson Tutorial Video* 11-7	☐ *Lesson Tutorial Video* 11-7
☐ Reading Strategies 11-7 CRB	☐ Problem Solving 11-7 CRB	☐ Problem Solving 11-7 CRB	☐ Reading Strategies 11-7 CRB
☐ *Questioning Strategies* pp. 194–195	☐ Multiple Representations TE p. 629	☐ Multiple Representations TE p. 629	
☐ *IDEA Works!* 11-7			☐ *Multilingual Glossary*

ASSESSMENT
- ☐ Lesson Quiz, TE p. 631 and DT 11-7   ☐ State-Specific Test Prep Online Keyword: MR7 TestPrep

Copyright © Holt, Rinehart and Winston.
All rights reserved.

Holt Mathematics

Teacher's Name _____ Class _____ Date _____

Lesson Plan 11-8
Solving Integer Equations pp. 636–639   Day _____

Objective Students solve equations containing integers.

> **NCTM Standards:** Understand meanings of operations and how they relate to one another; Use mathematical models to represent and understand quantitative relationships.

Pacing
☐ 45-minute Classes: 1 day   ☐ 90-minute Classes: 1/2 day   ☐ Other_____

WARM UP
☐ Warm Up TE p. 636 and Daily Transparency 11-8
☐ Problem of the Day TE p. 636 and Daily Transparency 11-8

TEACH
☐ Lesson Presentation CD-ROM 11-8
☐ Alternate Opener, Explorations Transparency 11-8, TE p. 636, and Exploration 11-8
☐ Reaching All Learners TE p. 637
☐ *Hands-On Lab Activities* 11-8
☐ *Know-It Notebook* 11-8

PRACTICE AND APPLY
☐ Example 1: Average: 1–8, 33–34, 43–45, 71–79 Advanced: 17–24, 33–34, 49–50, 71–79
☐ Example 2: Average: 1–50, 61–65, 71–79 Advanced: 17–79

REACHING ALL LEARNERS – Differentiated Instruction for students with

Developing Knowledge	On-level Knowledge	Advanced Knowledge	English Language Development
☐ Inclusion TE p. 637	☐ Cognitive Strategies TE p. 637	☐ Cognitive Strategies TE p. 637	☐ Cognitive Strategies TE p. 637
☐ Practice A 11-8 CRB	☐ Practice B 11-8 CRB	☐ Practice C 11-8 CRB	☐ Practice A, B, or C 11-8 CRB
☐ Reteach 11-8 CRB	☐ Puzzles, Twisters & Teasers 11-8 CRB	☐ Challenge 11-8 CRB	☐ *Success for ELL* 11-8
☐ Homework Help Online Keyword: MR7 11-8	☐ Homework Help Online Keyword: MR7 11-8	☐ Homework Help Online Keyword: MR7 11-8	☐ Homework Help Online Keyword: MR7 11-8
☐ *Lesson Tutorial Video* 11-8	☐ *Lesson Tutorial Video* 11-8	☐ *Lesson Tutorial Video* 11-8	☐ *Lesson Tutorial Video* 11-8
☐ Reading Strategies 11-8 CRB	☐ Problem Solving 11-8 CRB	☐ Problem Solving 11-8 CRB	☐ Reading Strategies 11-8 CRB
☐ *Questioning Strategies* pp. 196–197			
☐ *IDEA Works!* 11-8			☐ *Multilingual Glossary*

ASSESSMENT
☐ Lesson Quiz, TE p. 639 and DT 11-8   ☐ State-Specific Test Prep Online Keyword: MR7 TestPrep

Teacher's Name _____ Class _____ Date _____

Lesson Plan 11-9
Tables and Functions pp. 640–643   Day _____

Objective Students use data in a table to write an equation for a function and to use the equation to find a missing value.

> **NCTM Standards:** Understand patterns, relations, and functions; Use mathematical models to represent and understand quantitative relationships.

Pacing
- [ ] 45-minute Classes: 1 day
- [ ] 90-minute Classes: 1/2 day
- [ ] Other _____

WARM UP
- [ ] Warm Up TE p. 640 and Daily Transparency 11-9
- [ ] Problem of the Day TE p. 640 and Daily Transparency 11-9

TEACH
- [ ] Lesson Presentation CD-ROM 11-9
- [ ] Alternate Opener, Explorations Transparency 11-9, TE p. 640, and Exploration 11-9
- [ ] Reaching All Learners TE p. 641
- [ ] Teaching Transparency 11-9
- [ ] *Technology Lab Activities* 11-9
- [ ] *Know-It Notebook* 11-9

PRACTICE AND APPLY
- [ ] Example 1: Average: 1–2, 10–11, 22–30 Advanced: 5–6, 12–13, 22–30
- [ ] Example 2: Average: 1–3, 10–13, 22–30 Advanced: 5–8, 11–13, 22–30
- [ ] Example 3: Average: 1–11, 14–17, 22–30 Advanced: 5–14, 18–30

REACHING ALL LEARNERS – Differentiated Instruction for students with

Developing Knowledge	On-level Knowledge	Advanced Knowledge	English Language Development
☐ Multiple Representations TE p. 641	☐ Multiple Representations TE p. 641	☐ Multiple Representations TE p. 641	☐ Multiple Representations TE p. 641
☐ Practice A 11-9 CRB	☐ Practice B 11-9 CRB	☐ Practice C 11-9 CRB	☐ Practice A, B, or C 11-9 CRB
☐ Reteach 11-9 CRB	☐ Puzzles, Twisters & Teasers 11-9 CRB	☐ Challenge 11-9 CRB	☐ *Success for ELL* 11-9
☐ Homework Help Online Keyword: MR7 11-9	☐ Homework Help Online Keyword: MR7 11-9	☐ Homework Help Online Keyword: MR7 11-9	☐ Homework Help Online Keyword: MR7 11-9
☐ *Lesson Tutorial Video* 11-9	☐ *Lesson Tutorial Video* 11-9	☐ *Lesson Tutorial Video* 11-9	☐ *Lesson Tutorial Video* 11-9
☐ Reading Strategies 11-9 CRB	☐ Problem Solving 11-9 CRB	☐ Problem Solving 11-9 CRB	☐ Reading Strategies 11-9 CRB
☐ *Questioning Strategies* pp. 198–199			☐ Lesson Vocabulary SE p. 640
☐ *IDEA Works!* 11-9			☐ *Multilingual Glossary*

ASSESSMENT
- [ ] Lesson Quiz, TE p. 643 and DT 11-9
- [ ] State-Specific Test Prep Online Keyword: MR7 TestPrep

Teacher's Name _____ Class _____ Date _____

Lesson Plan 11-10
Graphing Functions pp. 646–649   Day _____

Objective Students represent linear functions using ordered pairs and graphs.

> **NCTM Standards:** Understand patterns, relations, and functions; Use mathematical models to represent and understand quantitative relationships.

Pacing
☐ 45-minute Classes: 1 day   ☐ 90-minute Classes: 1/2 day   ☐ Other_____

WARM UP
☐ Warm Up TE p. 646 and Daily Transparency 11-10
☐ Problem of the Day TE p. 646 and Daily Transparency 11-10

TEACH
☐ Lesson Presentation CD-ROM 11-10
☐ Alternate Opener, Explorations Transparency 11-10, TE p. 646, and Exploration 11-10
☐ Reaching All Learners TE p. 647
☐ *Hands-On Lab Activities* 11-10
☐ *Technology Lab Activities* 11-10
☐ *Know-It Notebook* 11-10

PRACTICE AND APPLY
☐ Example 1: Average: 1–2, 38–46 Advanced: 11–12, 38–46
☐ Example 2: Average: 1–4, 29, 38–46 Advanced: 11–14, 29, 38–46
☐ Example 3: Average: 1–7, 29, 38–46 Advanced: 11–20, 29, 38–46
☐ Example 4: Average: 1–18, 24–33, 38–46 Advanced: 11–46

REACHING ALL LEARNERS – Differentiated Instruction for students with

Developing Knowledge	On-level Knowledge	Advanced Knowledge	English Language Development
☐ Cooperative Learning TE p. 647	☐ Cooperative Learning TE p. 647	☐ Cooperative Learning TE p. 647	☐ Cooperative Learning TE p. 647
☐ Practice A 11-10 CRB	☐ Practice B 11-10 CRB	☐ Practice C 11-10 CRB	☐ Practice A, B, or C 11-10 CRB
☐ Reteach 11-10 CRB	☐ Puzzles, Twisters & Teasers 11-10 CRB	☐ Challenge 11-10 CRB	☐ *Success for ELL* 11-10
☐ Homework Help Online Keyword: MR7 11-10	☐ Homework Help Online Keyword: MR7 11-10	☐ Homework Help Online Keyword: MR7 11-10	☐ Homework Help Online Keyword: MR7 11-10
☐ *Lesson Tutorial Video* 11-10	☐ *Lesson Tutorial Video* 11-10	☐ *Lesson Tutorial Video* 11-10	☐ *Lesson Tutorial Video* 11-10
☐ Reading Strategies 11-10 CRB	☐ Problem Solving 11-10 CRB	☐ Problem Solving 11-10 CRB	☐ Reading Strategies 11-10 CRB
☐ *Questioning Strategies* pp. 200–201	☐ Modeling TE p. 647	☐ Modeling TE p. 647	☐ Lesson Vocabulary SE p. 646
☐ *IDEA Works!* 11-10			☐ *Multilingual Glossary*

ASSESSMENT
☐ Lesson Quiz, TE p. 649 and DT 11-10   ☐ State-Specific Test Prep Online Keyword: MR7 TestPrep

Teacher's Name _____ Class _____ Date _____

Lesson Plan 12-1
Introduction to Probability pp. 668–671   Day _____

Objective Students estimate the likelihood of an event and write and compare probabilities.

> **NCTM Standards:** Understand and apply basic concepts of probability; Select and use various types of reasoning and methods of proof; Recognize and apply mathematics in contexts outside of mathematics.

Pacing
☐ 45-minute Classes: 1 day   ☐ 90-minute Classes: 1/2 day   ☐ Other_____

WARM UP
☐ Warm Up TE p. 668 and Daily Transparency 12-1
☐ Problem of the Day TE p. 668 and Daily Transparency 12-1

TEACH
☐ Lesson Presentation CD-ROM 12-1
☐ Alternate Opener, Explorations Transparency 12-1, TE p. 668, and Exploration 12-1
☐ Reaching All Learners TE p. 669
☐ Teaching Transparency 12-1
☐ *Know-It Notebook* 12-1

PRACTICE AND APPLY
☐ Example 1: Average: 1–2, 12–18, 23–32 Advanced: 5–8, 14–18, 23–32
☐ Example 2: Average: 1–3, 12–18, 23–32 Advanced: 5–9, 15–18, 23–32
☐ Example 3: Average: 1–11, 16–32 Advanced: 5–32

REACHING ALL LEARNERS – Differentiated Instruction for students with

Developing Knowledge	On-level Knowledge	Advanced Knowledge	English Language Development
☐ Inclusion TE p. 669	☐ Cooperative Learning TE p. 669	☐ Cooperative Learning TE p. 669	☐ Cooperative Learning TE p. 669
☐ Practice A 12-1 CRB	☐ Practice B 12-1 CRB	☐ Practice C 12-1 CRB	☐ Practice A, B, or C 12-1 CRB
☐ Reteach 12-1 CRB	☐ Puzzles, Twisters & Teasers 12-1 CRB	☐ Challenge 12-1 CRB	☐ *Success for ELL* 12-1
☐ Homework Help Online Keyword: MR7 12-1	☐ Homework Help Online Keyword: MR7 12-1	☐ Homework Help Online Keyword: MR7 12-1	☐ Homework Help Online Keyword: MR7 12-1
☐ *Lesson Tutorial Video* 12-1	☐ *Lesson Tutorial Video* 12-1	☐ *Lesson Tutorial Video* 12-1	☐ *Lesson Tutorial Video* 12-1
☐ Reading Strategies 12-1 CRB	☐ Problem Solving 12-1 CRB	☐ Problem Solving 12-1 CRB	☐ Reading Strategies 12-1 CRB
☐ *Questioning Strategies* pp. 202–203			☐ Lesson Vocabulary SE p. 668
☐ *IDEA Works!* 12-1			☐ *Multilingual Glossary*

ASSESSMENT
☐ Lesson Quiz, TE p. 671 and DT 12-1   ☐ State-Specific Test Prep Online Keyword: MR7 TestPrep

Teacher's Name _____ Class _____ Date _____

Lesson Plan 12-2
Experimental Probability pp. 672–675 Day _____

Objective Students find the experimental probability of an event.

> **NCTM Standards:** Understand and apply basic concepts of probability; Select and use various types of reasoning and methods of proof.

Pacing
☐ 45-minute Classes: 1 day ☐ 90-minute Classes: 1/2 day ☐ Other_____

WARM UP
☐ Warm Up TE p. 672 and Daily Transparency 12-2
☐ Problem of the Day TE p. 672 and Daily Transparency 12-2

TEACH
☐ Lesson Presentation CD-ROM 12-2
☐ Alternate Opener, Explorations Transparency 12-2, TE p. 672, and Exploration 12-2
☐ Reaching All Learners TE p. 673
☐ Teaching Transparency 12-2
☐ *Technology Lab Activities* 12-2
☐ *Know-It Notebook* 12-2

PRACTICE AND APPLY
☐ Example 1: Average: 1, 9–10, 15–22 Advanced: 4–5, 10, 15–22
☐ Example 2: Average: 1–2, 9–11, 15–22 Advanced: 4–7, 11, 15–22
☐ Example 3: Average: 1–8, 12–22 Advanced: 4–22

REACHING ALL LEARNERS – Differentiated Instruction for students with

Developing Knowledge	On-level Knowledge	Advanced Knowledge	English Language Development
☐ Kinesthetic Experience TE p. 673	☐ Kinesthetic Experience TE p. 673	☐ Kinesthetic Experience TE p. 673	☐ Kinesthetic Experience TE p. 673
☐ Practice A 12-2 CRB	☐ Practice B 12-2 CRB	☐ Practice C 12-2 CRB	☐ Practice A, B, or C 12-2 CRB
☐ Reteach 12-2 CRB	☐ Puzzles, Twisters & Teasers 12-2 CRB	☐ Challenge 12-2 CRB	☐ *Success for ELL* 12-2
☐ Homework Help Online Keyword: MR7 12-2	☐ Homework Help Online Keyword: MR7 12-2	☐ Homework Help Online Keyword: MR7 12-2	☐ Homework Help Online Keyword: MR7 12-2
☐ *Lesson Tutorial Video* 12-2	☐ *Lesson Tutorial Video* 12-2	☐ *Lesson Tutorial Video* 12-2	☐ *Lesson Tutorial Video* 12-2
☐ *Reading Strategies* 12-2 CRB	☐ *Problem Solving* 12-2 CRB	☐ *Problem Solving* 12-2 CRB	☐ *Reading Strategies* 12-2 CRB
☐ *Questioning Strategies* pp. 204–205			☐ Lesson Vocabulary SE p. 672
☐ *IDEA Works!* 12-2			☐ *Multilingual Glossary*

ASSESSMENT
☐ Lesson Quiz, TE p. 675 and DT 12-2 ☐ State-Specific Test Prep Online Keyword: MR7 TestPrep

Teacher's Name _____ Class _____ Date _____

Lesson Plan 12-3
Counting Methods and Sample Spaces pp. 678–681   Day _____

Objective Students use counting methods to find all possible outcomes.

> **NCTM Standards:** Compute fluently and make reasonable estimates; Develop and evaluate mathematical arguments and proofs; Recognize reasoning and proof as fundamental aspects of mathematics.

Pacing
☐ 45-minute Classes: 1 day   ☐ 90-minute Classes: 1/2 day   ☐ Other_____

WARM UP
☐ Warm Up TE p. 678 and Daily Transparency 12-3
☐ Problem of the Day TE p. 678 and Daily Transparency 12-3

TEACH
☐ Lesson Presentation CD-ROM 12-3
☐ Alternate Opener, Explorations Transparency 12-3, TE p. 678, and Exploration 12-3
☐ Reaching All Learners TE p. 679
☐ *Hands-On Lab Activities* 12-3
☐ *Know-It Notebook* 12-3

PRACTICE AND APPLY
☐ Example 1: Average: 1, 7–9, 13–19 Advanced: 4, 7–9, 13–19
☐ Example 2: Average: 1–2, 7–9, 13–19 Advanced: 4–5, 7–9, 13–19
☐ Example 3: Average: 1–7, 10–19 Advanced: 4–19

REACHING ALL LEARNERS – Differentiated Instruction for students with

Developing Knowledge	On-level Knowledge	Advanced Knowledge	English Language Development
☐ Diversity TE p. 679	☐ Diversity TE p. 679	☐ Diversity TE p. 679	☐ Diversity TE p. 679
☐ Practice A 12-3 CRB	☐ Practice B 12-3 CRB	☐ Practice C 12-3 CRB	☐ Practice A, B, or C 12-3 CRB
☐ Reteach 12-3 CRB	☐ Puzzles, Twisters & Teasers 12-3 CRB	☐ Challenge 12-3 CRB	☐ *Success for ELL* 12-3
☐ Homework Help Online Keyword: MR7 12-3	☐ Homework Help Online Keyword: MR7 12-3	☐ Homework Help Online Keyword: MR7 12-3	☐ Homework Help Online Keyword: MR7 12-3
☐ *Lesson Tutorial Video* 12-3	☐ *Lesson Tutorial Video* 12-3	☐ *Lesson Tutorial Video* 12-3	☐ *Lesson Tutorial Video* 12-3
☐ Reading Strategies 12-3 CRB	☐ Problem Solving 12-3 CRB	☐ Problem Solving 12-3 CRB	☐ Reading Strategies 12-3 CRB
☐ *Questioning Strategies* pp. 206–207	☐ Critical Thinking TE p. 679	☐ Critical Thinking TE p. 679	☐ Lesson Vocabulary SE p. 678
☐ *IDEA Works!* 12-3			☐ *Multilingual Glossary*

ASSESSMENT
☐ Lesson Quiz, TE p. 681 and DT 12-3   ☐ State-Specific Test Prep Online Keyword: MR7 TestPrep

Teacher's Name _____ Class _____ Date _____

Lesson Plan 12-4
Theoretical Probability pp. 682–685 Day _____

Objective Students find the theoretical probability and complement of an event.

> **NCTM Standards:** Understand and apply basic concepts of probability; Select and use various types of reasoning and methods of proof.

Pacing
☐ 45-minute Classes: 1 day   ☐ 90-minute Classes: 1/2 day   ☐ Other _____

WARM UP
☐ Warm Up TE p. 682 and Daily Transparency 12-4
☐ Problem of the Day TE p. 682 and Daily Transparency 12-4

TEACH
☐ Lesson Presentation CD-ROM 12-4
☐ Alternate Opener, Explorations Transparency 12-4, TE p. 682, and Exploration 12-4
☐ Reaching All Learners TE p. 683
☐ Teaching Transparency 12-4
☐ *Hands-On Lab Activities* 12-4
☐ *Know-It Notebook* 12-4

PRACTICE AND APPLY
☐ Example 1: Average: 1–2, 10–17, 37–43 Advanced: 5–7, 10–17, 37–43
☐ Example 2: Average: 1–13, 18–43 Advanced: 5–43

REACHING ALL LEARNERS – Differentiated Instruction for students with

Developing Knowledge	On-level Knowledge	Advanced Knowledge	English Language Development
☐ Curriculum Integration TE p. 683	☐ Curriculum Integration TE p. 683	☐ Curriculum Integration TE p. 683	☐ Curriculum Integration TE p. 683
☐ Practice A 12-4 CRB	☐ Practice B 12-4 CRB	☐ Practice C 12-4 CRB	☐ Practice A, B, or C 12-4 CRB
☐ Reteach 12-4 CRB	☐ Puzzles, Twisters & Teasers 12-4 CRB	☐ Challenge 12-4 CRB	☐ *Success for ELL* 12-4
☐ Homework Help Online Keyword: MR7 12-4	☐ Homework Help Online Keyword: MR7 12-4	☐ Homework Help Online Keyword: MR7 12-4	☐ Homework Help Online Keyword: MR7 12-4
☐ *Lesson Tutorial Video* 12-4	☐ *Lesson Tutorial Video* 12-4	☐ *Lesson Tutorial Video* 12-4	☐ *Lesson Tutorial Video* 12-4
☐ Reading Strategies 12-4 CRB	☐ Problem Solving 12-4 CRB	☐ Problem Solving 12-4 CRB	☐ Reading Strategies 12-4 CRB
☐ *Questioning Strategies* pp. 208–209			☐ Lesson Vocabulary SE p. 682
☐ *IDEA Works!* 12-4			☐ *Multilingual Glossary*

ASSESSMENT
☐ Lesson Quiz, TE p. 685 and DT 12-4   ☐ State-Specific Test Prep Online Keyword: MR7 TestPrep

Teacher's Name _____ Class _____ Date _____

Lesson Plan 12-5
Compound Events pp. 688–691   Day _____

Objective Students list all the outcomes and find the theoretical probability of a compound event.

NCTM Standards: Understand and apply basic concepts of probability.

Pacing
☐ 45-minute Classes: 1 day   ☐ 90-minute Classes: 1/2 day   ☐ Other _____

WARM UP
☐ Warm Up TE p. 688 and Daily Transparency 12-5
☐ Problem of the Day TE p. 688 and Daily Transparency 12-5

TEACH
☐ Lesson Presentation CD-ROM 12-5
☐ Alternate Opener, Explorations Transparency 12-5, TE p. 688, and Exploration 12-5
☐ Reaching All Learners TE p. 689
☐ *Technology Lab Activities* 12-5
☐ *Know-It Notebook* 12-5

PRACTICE AND APPLY
☐ Example 1: Average: 1–11, 14–29 Advanced: 3–29

REACHING ALL LEARNERS – Differentiated Instruction for students with

Developing Knowledge	On-level Knowledge	Advanced Knowledge	English Language Development
☐ Concrete Manipulatives TE p. 689	☐ Concrete Manipulatives TE p. 689	☐ Concrete Manipulatives TE p. 689	☐ Concrete Manipulatives TE p. 689
☐ Practice A 12-5 CRB	☐ Practice B 12-5 CRB	☐ Practice C 12-5 CRB	☐ Practice A, B, or C 12-5 CRB
☐ Reteach 12-5 CRB	☐ Puzzles, Twisters & Teasers 12-5 CRB	☐ Challenge 12-5 CRB	☐ *Success for ELL* 12-5
☐ Homework Help Online Keyword: MR7 12-5	☐ Homework Help Online Keyword: MR7 12-5	☐ Homework Help Online Keyword: MR7 12-5	☐ Homework Help Online Keyword: MR7 12-5
☐ *Lesson Tutorial Video* 12-5	☐ *Lesson Tutorial Video* 12-5	☐ *Lesson Tutorial Video* 12-5	☐ *Lesson Tutorial Video* 12-5
☐ Reading Strategies 12-5 CRB	☐ Problem Solving 12-5 CRB	☐ Problem Solving 12-5 CRB	☐ Reading Strategies 12-5 CRB
☐ *Questioning Strategies* pp. 210–211	☐ Multiple Representations TE p. 689	☐ Multiple Representations TE p. 689	☐ Lesson Vocabulary SE p. 688
☐ *IDEA Works!* 12-5			☐ *Multilingual Glossary*

ASSESSMENT
☐ Lesson Quiz, TE p. 691 and DT 12-5   ☐ State-Specific Test Prep Online Keyword: MR7 TestPrep

Holt Mathematics

Teacher's Name _____ Class _____ Date _____

Lesson Plan 12-6
Making Predictions pp. 694–697   Day _____

Objective Students use probability to predict future events.

> **NCTM Standards:** Develop and evaluate inferences and predictions that are based on data; Understand and apply basic concepts of probability; Recognize and apply mathematics in contexts outside of mathematics.

Pacing
☐ 45-minute Classes: 1 day   ☐ 90-minute Classes: 1/2 day   ☐ Other_____

WARM UP
☐ Warm Up TE p. 694 and Daily Transparency 12-6
☐ Problem of the Day TE p. 694 and Daily Transparency 12-6

TEACH
☐ Lesson Presentation CD-ROM 12-6
☐ Alternate Opener, Explorations Transparency 12-6, TE p. 694, and Exploration 12-6
☐ Reaching All Learners TE p. 695
☐ *Know-It Notebook* 12-6

PRACTICE AND APPLY
☐ Example 1: Average: 1, 8, 15–24 Advanced: 4, 8, 15–24
☐ Example 2: Average: 1–2, 8, 15–24 Advanced: 4–6, 8, 15–24
☐ Example 3: Average: 1–7, 10–24 Advanced: 4–24

REACHING ALL LEARNERS – Differentiated Instruction for students with

Developing Knowledge	On-level Knowledge	Advanced Knowledge	English Language Development
☐ Cooperative Learning TE p. 695	☐ Cooperative Learning TE p. 695	☐ Cooperative Learning TE p. 695	☐ Cooperative Learning TE p. 695
☐ Practice A 12-6 CRB	☐ Practice B 12-6 CRB	☐ Practice C 12-6 CRB	☐ Practice A, B, or C 12-6 CRB
☐ Reteach 12-6 CRB	☐ Puzzles, Twisters & Teasers 12-6 CRB	☐ Challenge 12-6 CRB	☐ *Success for ELL* 12-6
☐ Homework Help Online Keyword: MR7 12-6	☐ Homework Help Online Keyword: MR7 12-6	☐ Homework Help Online Keyword: MR7 12-6	☐ Homework Help Online Keyword: MR7 12-6
☐ *Lesson Tutorial Video* 12-6	☐ *Lesson Tutorial Video* 12-6	☐ *Lesson Tutorial Video* 12-6	☐ *Lesson Tutorial Video* 12-6
☐ Reading Strategies 12-6 CRB	☐ Problem Solving 12-6 CRB	☐ Problem Solving 12-6 CRB	☐ Reading Strategies 12-6 CRB
☐ *Questioning Strategies* pp. 212–213			☐ Lesson Vocabulary SE p. 694
☐ *IDEA Works!* 12-6			☐ *Multilingual Glossary*

ASSESSMENT
☐ Lesson Quiz, TE p. 697 and DT 12-6   ☐ State-Specific Test Prep Online Keyword: MR7 TestPrep